RSLogix 5000 PLC Lab Manual for use with

Programmable Logic Controllers

Frank D. Petruzella

RSLogix 5000 PLC Lab Manual for use with

Programmable Logic Controllers

Sixth Edition

McGraw Hill

McGraw Hill

RSLOGIX 5000 PLC LAB MANUAL FOR USE WITH PROGRAMMABLE LOGIC CONTROLLERS, SIXTH EDITION

Published by McGraw Hill LLC, 1325 Avenue of the Americas, New York, NY 10019.

This book is printed on acid-free paper.

1 2 3 4 5 6 7 8 9 LKV 27 26 25 24 23 22

ISBN 978-1-264-81658-3
MHID 1-264-81658-8

Cover image: *Xmentoys/Shutterstock*

mheducation.com/highered

Contents

NOTE: *Chapter 14: Process Control, Network Systems, and SCADA* and *Chapter 15: ControlLogix Controllers* contain no assignments.

Preface

This lab manual contains more than 180 programming exercises and may be used with Rockwell's RSLogix 5000 or PLCLogix simulation software. The programming exercises are designed to be used in parallel with *Programmable Logic Controllers*, 6/e, to provide a hands-on learning experience. The exercises cover a broad range of PLC programming topics including bit instructions, timers, counters, math, sequencers, shift register, data manipulation, program control, and troubleshooting.

Frank D. Petruzella

About the Author

Frank D. Petruzella has extensive practical experience in the electrical control field, as well as many years of experience teaching and authoring textbooks. Before becoming a full-time educator, he was employed as an apprentice and electrician in areas of electrical installation and maintenance. He holds a Master of Science degree from Niagara University, a Bachelor of Science degree from the State University of New York College–Buffalo, as well as diplomas in Electrical Power and Electronics from the Erie County Technical Institute.

Getting Started with Logix 5000

Getting Started with Logix 5000

PLCLogix is a PLC emulator that simulates the RSLogix 5000 (rev. 17) programming environment using a 1756-L63 ControlLogix controller in a 10-Slot chassis. It allows the user to practice programming using identical instructions and simulated I/O to solve real-world type problems.

Creating a File

The first step in creating any new program is to a make a new file. These files are usually controller specific, because all programming instructions are not necessarily available for all controllers. In PLCLogix, the controller is not selectable, but a new file name and location must be selected. The file name must start with a letter, contain no spaces, no special characters, and not exceed length limits which vary by controller. When using actual hardware, care must be taken to select the exact controller that is being used, and the correct firmware version, otherwise a major error will occur.

The New Controller window (shown below) is accessed by clicking on the file menu, and then clicking New.

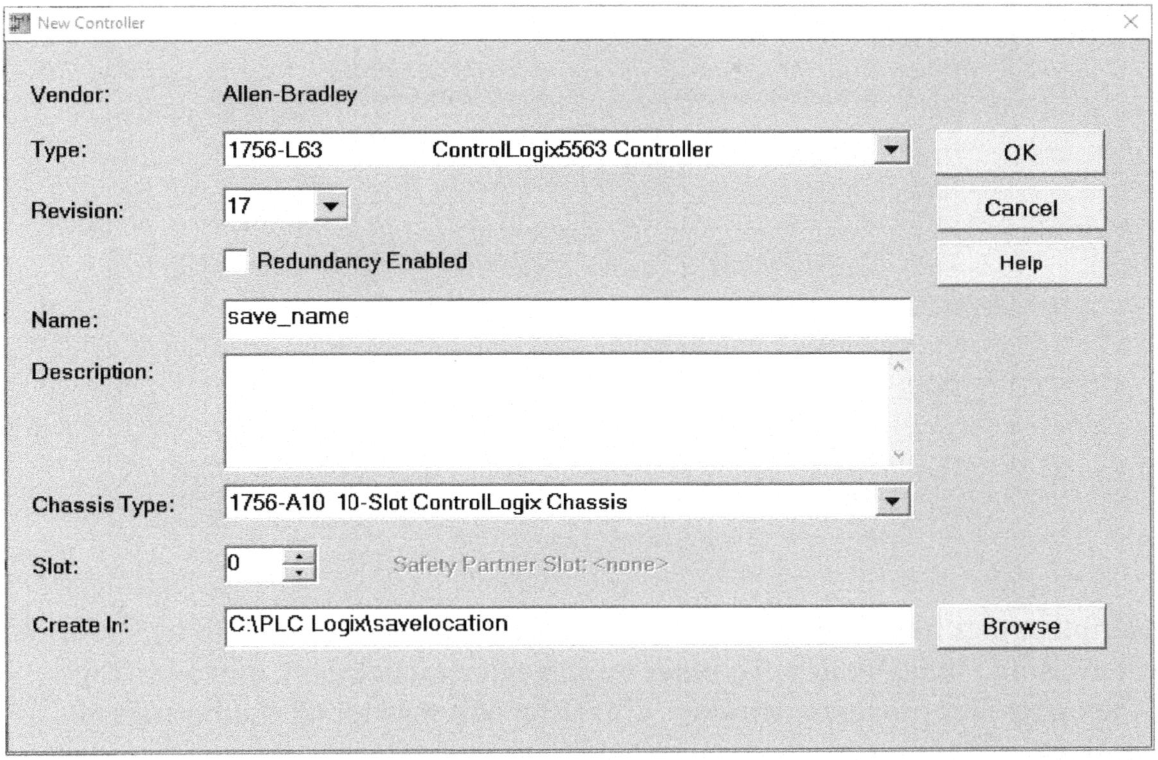

Creating Tags

The next step is to create tags. In Logix 5000, base tags are automatically created when I/O hardware is added. To add a new module, Right click on I/O configuration in the controller organizer window, and click on new module. It is extremely important that the I/O configuration in the program is identical to the actual real-world configuration. In PLCLogix, the I/O hardware has already been added and cannot be changed.

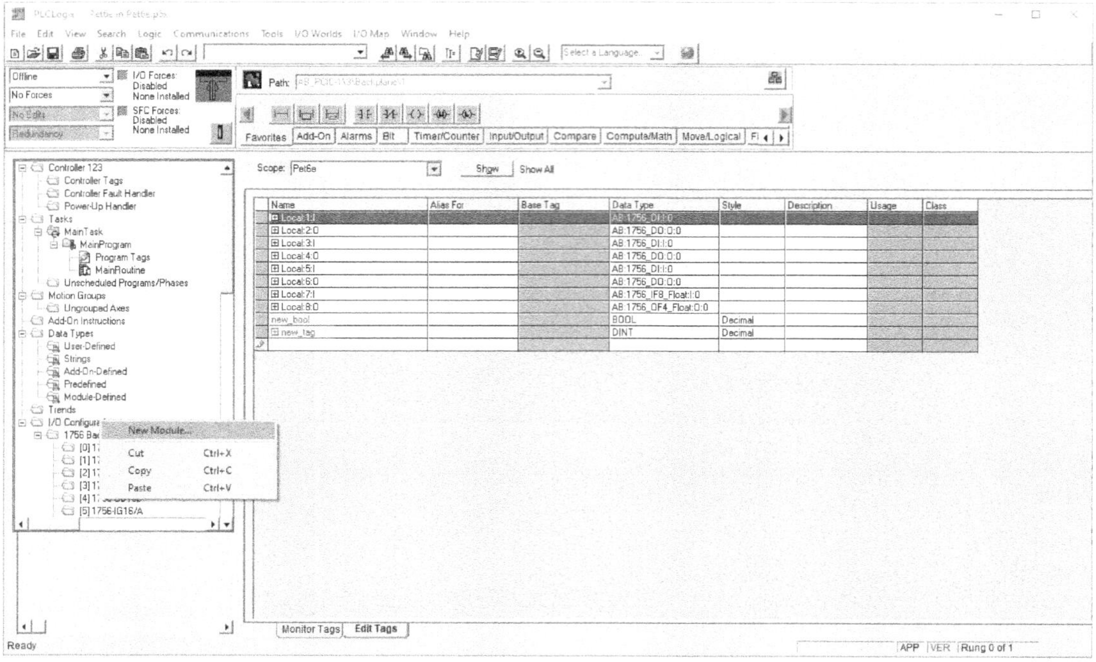

Descriptive tags like the ones highlighted in the window above, can be created using the tag editor. To enter the tag editor, click on Controller Tags on the controller organizer window. Click the edit tags tab at the bottom of the screen, and type a new tag name next to the pencil icon. Be sure to select the correct data type such as BOOL for a bit tag, DINT for an integer tag, and REAL for a floating point decimal tag.

Note: it is possible to start programming ladder logic using question marks "?" for tags, and filling in the tag names later.

Creating a Program

Once tags have been created, it is time to program. The program structure for a Logix 5000 controller is Task, Program, and Routine. By default, a Main task, Main program, and Main routine are created in the project. Ladder logic can be programmed directly in the Main routine, or written in subroutines that are called in the Main routine using the JSR instruction. The Main routine is found in the controller organizer by first expanding Tasks, then MainTask, then MainProgram, and finally double clicking MainRoutine to open it.

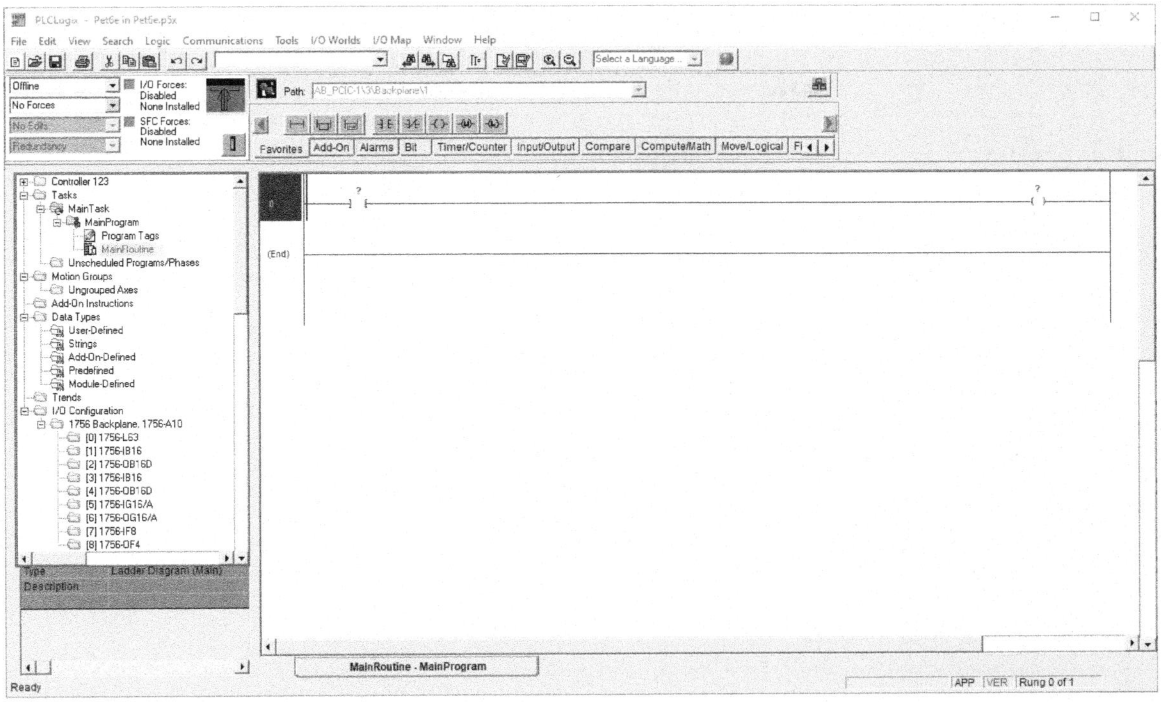

To begin programming, drag an instruction from the menu onto a rung. Click the ? above the instruction and type in the desired tag name, or select one from the drop down menu. Note the tag data type must match the instruction data type, BOOL for a bit instruction, etc. Local:1:I.Data, for example, is a DINT type and won't work for an XIC, which is a bit instruction. Local:1:I.Data.0 is one bit of Local:1:I.Data, is a BOOL data type, and can be used by an XIC instruction to read an input.

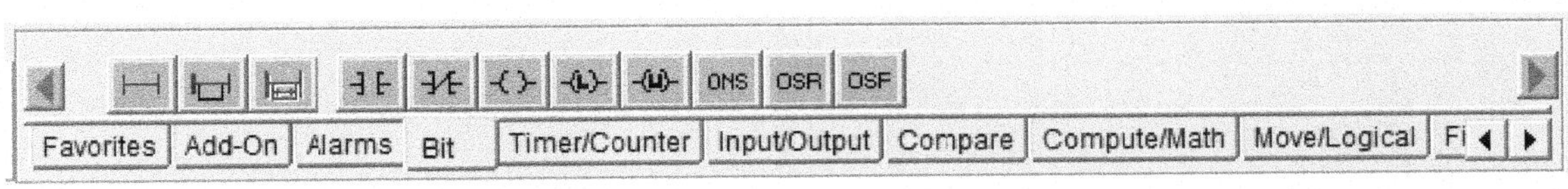

To create a new rung, click the rung icon 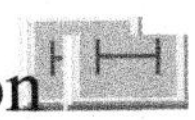 on the instruction menu, or right click in the left margin of the ladder diagram and select insert rung from the menu. To add a branch to a rung, select the rung to branch, and click the branch icon  The branch will appear on the rung. Instructions can be moved in or out of the branch by dragging them.

0
Local:1:I.Data.0
?
?
1
?
?
(End)

Testing a Program

To test the program in PLCLogix, it is not necessary to download the program. Go online by selecting Go Online from the communications menu, or by selecting Online from the dropdown above the controller organizer.

PLCLogix includes virtual I/O worlds. The I/O worlds can be selected from the top menu. The I/O rack interface is pictured below. Slots 1 and 3 simulate discrete inputs. The input type can be toggled between NO switch, NO limit switch, NC limit switch, NC pushbutton, and NO pushbutton by right clicking on the switch symbol. Slots 2 and 4 simulate discrete outputs. The indicator to the left of the address turns red when the output is on. Slot 5 is a BCD input simulator, its value is changed by clicking the up or down arrows on the thumbwheel switches. Slot 6 is a BCD output simulator with a 4 digit seven-segment display. Slot 7 is an analog input simulator. The input type can be toggled by right clicking on the symbol below the slider. The options are voltage, amperage, resistance, and temperature. The slider controls the input value. Slot 8 is an analog output simulator. The output type can be toggled by right clicking on the schematic symbol above the numerical display.

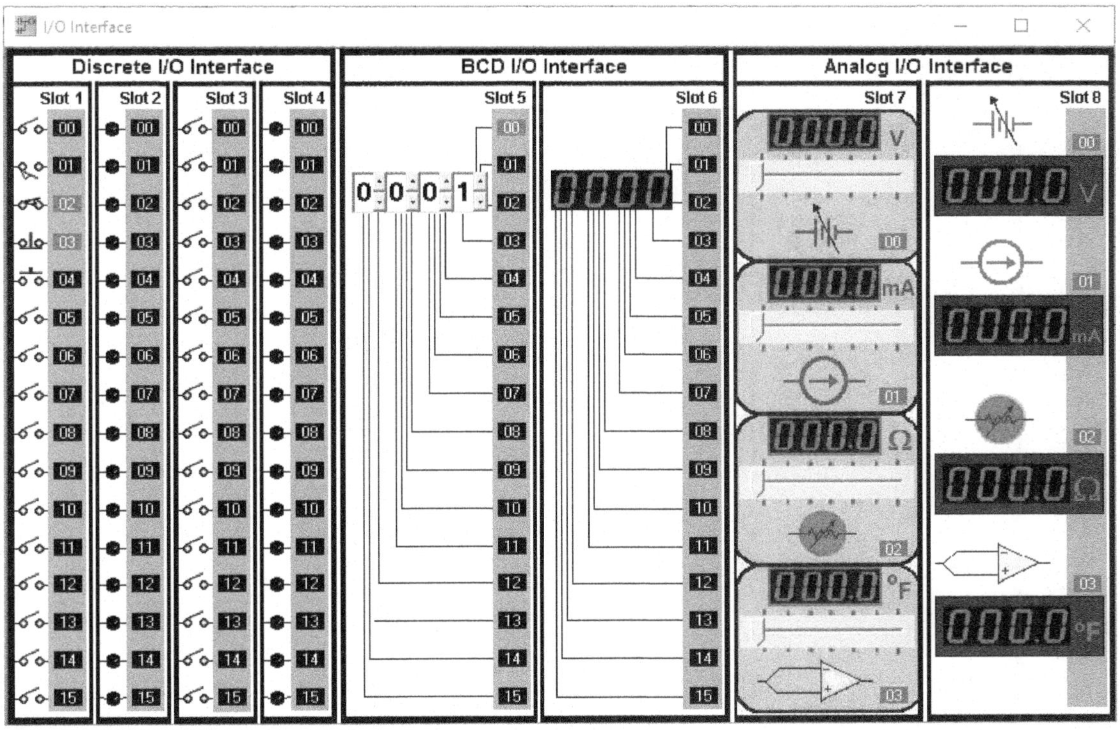

PLCLogix 5000 Simulator

This simulator **is not** included in the Manual and must be purchased separately. Visit **https://www.logicdesign.com/plclogix_5000.php** to purchase and download the software. This site contains all the information regarding the features of the program including information. Note that a 50% student discount is available for students currently enrolled in a PLC course which uses the accompanying textbook.

Acknowledgments

Gary Richardson Logic Design Inc. for help in working with his colleagues in getting this project started and seeing it through to completion.

Dwayne Nehls Program Simulator Consultant for technical editing and providing all of the solved simulation programming assignments.

NAME ____________________________________ DATE_______________

CHAPTER 1

Programmable Logic Controllers (PLCs): An Overview

1-1(a) Write a program that will execute the relay control circuit shown.

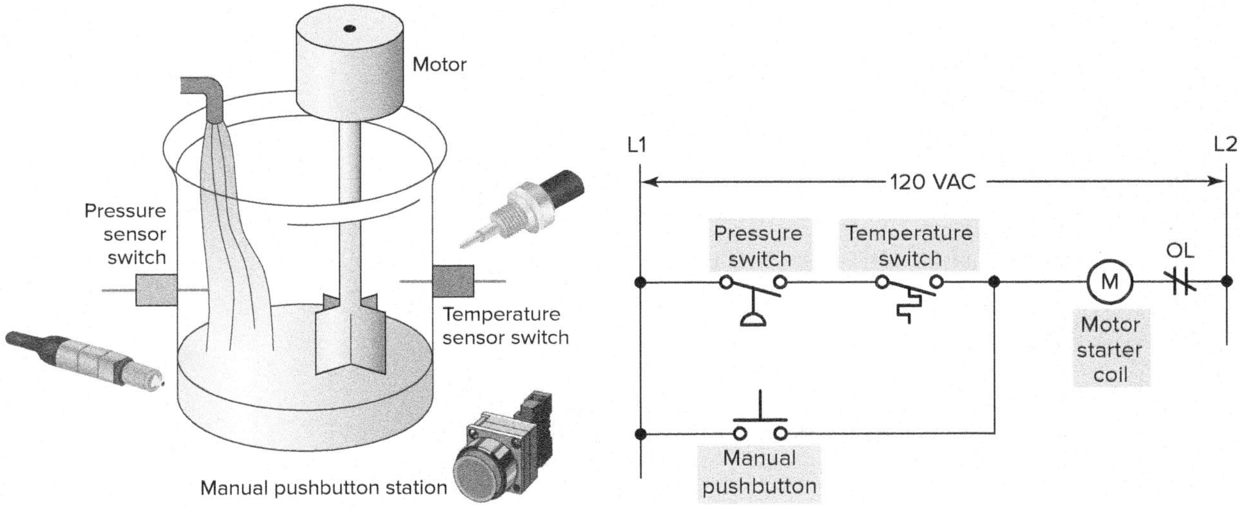

1-1(b) Modify the original program so that operating the pushbutton will turn the motor on only if the temperature switch is closed.

1-1(c) Modify the original program to operate so that all three input devices must be closed for the motor to operate.

1-1(d) Modify the original program to operate so that closing any one of the three input devices will turn the motor on.

1-1(e) Assume you have been required to install a main process On/Off switch to the original program. Write a program that will accommodate the request.

1-2 Given two single pole switches, write a program that will turn a pilot light on when both switch A and switch B are closed.

1-3 Given two normally open pushbuttons, write a program that will turn a pilot light on when either pushbutton A or pushbutton B is closed.

NAME ______________________________ DATE______________

1-4 Given four single pole switches (A-B-C-D), write a program that will turn on a pilot light if switches A and B or C and D are closed.

1-5 Write a program that will execute the relay control circuit shown.

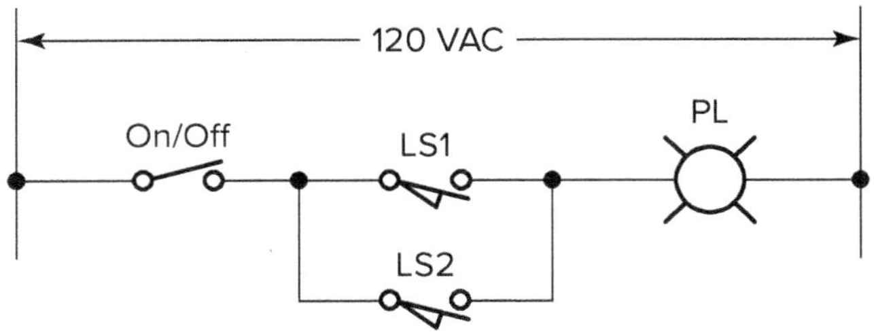

1-6 Write a program that will execute the relay control circuit shown.

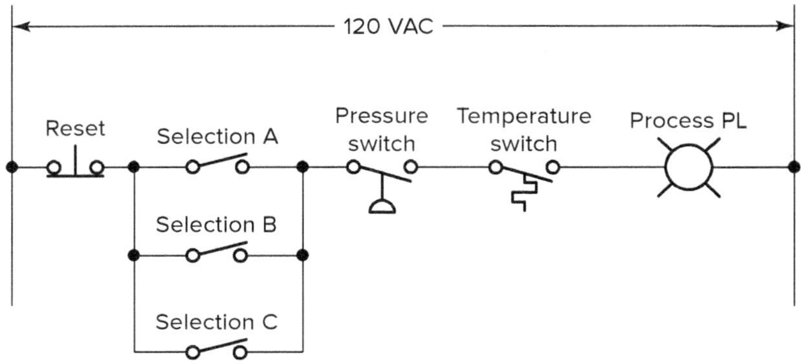

NAME ________________________________ DATE______________

CHAPTER 2

PLC Hardware Components

2-1(a) Construct a program which will emulate the 3-wire motor control program shown.

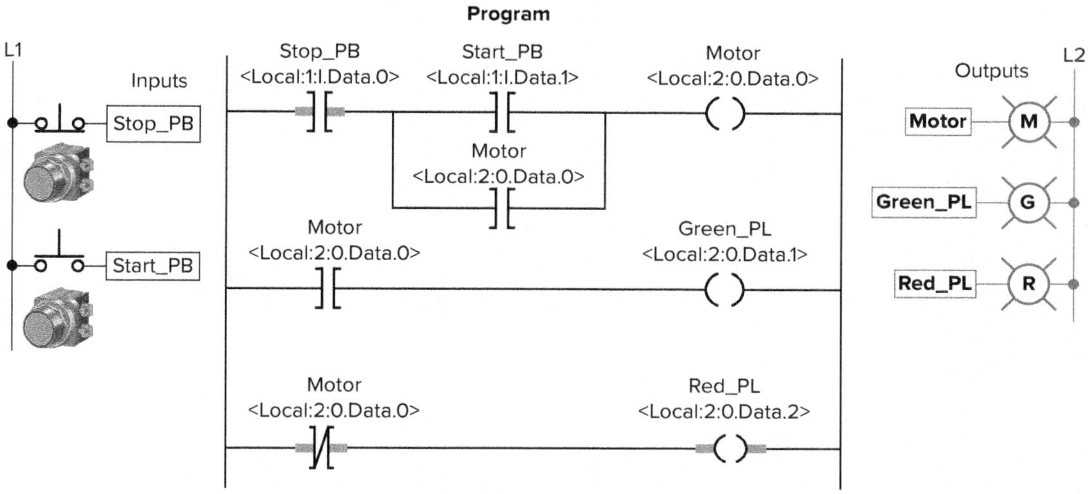

2-1(b) Modify the program for 2-wire control, so that the motor operates only as long as the Start_PB is held closed.

2-1(c) Assume you have been required to install a second Start/Stop station to the original 3-wire program. Write a program that will accomplish this task.

NAME ______________________________ DATE______________

2-2 Construct a program which will emulate the input memory bit status for the four types of inputs shown. Individually operate each of the input devices and record the value of the corresponding memory cell value (0 or 1) for each open and closed state.

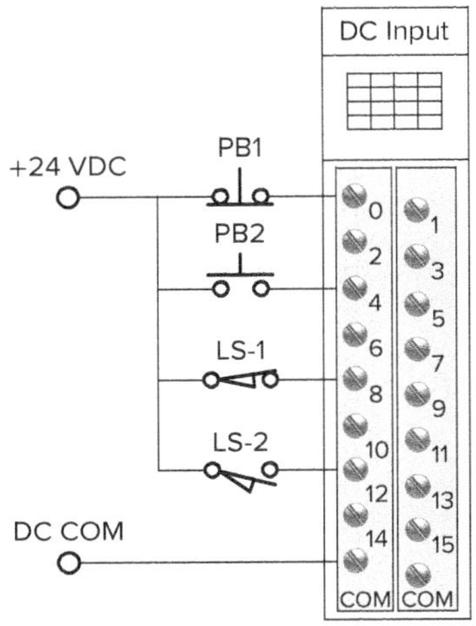

Slot 1 Tag monitor

15	14	13	12	11	10	9	8	7	6	5	4	3	2	1	0	Terminals
0	0	0	0	0	0	0	0	0	0	0	0	0	0	0	0	Bits
0	0	0	0	0	0	0	0	0	0	0	0	0	0	0	0	
0	0	0	0	0	0	0	0	0	0	0	0	0	0	0	0	
0	0	0	0	0	0	0	0	0	0	0	0	0	0	0	0	
0	0	0	0	0	0	0	0	0	0	0	0	0	0	0	0	

2-3 Construct a program which will emulate the output memory bit status for the output module devices shown. Each output is to be operated by its associated input On/Off switch. Individually operate each device on and off and record the value of the corresponding memory cell (0 or 1) state.

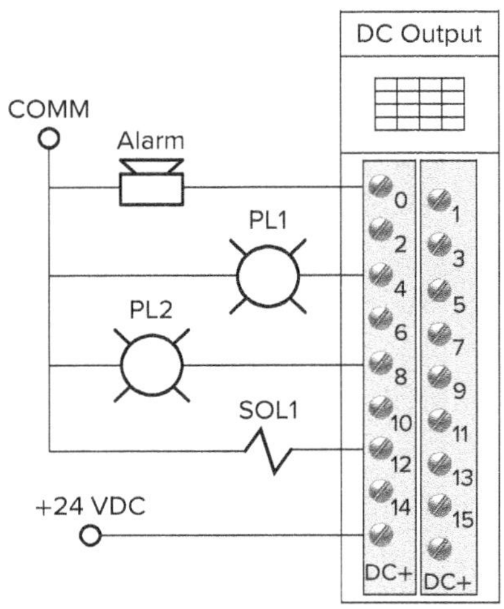

Slot 2 Tag monitor

15	14	13	12	11	10	9	8	7	6	5	4	3	2	1	0	Terminals
0	0	0	0	0	0	0	0	0	0	0	0	0	0	0	0	Bits
0	0	0	0	0	0	0	0	0	0	0	0	0	0	0	0	
0	0	0	0	0	0	0	0	0	0	0	0	0	0	0	0	
0	0	0	0	0	0	0	0	0	0	0	0	0	0	0	0	
0	0	0	0	0	0	0	0	0	0	0	0	0	0	0	0	

NAME ______________________________ DATE______________

2-4 Write a program for the 3-wire two motor starters (M1 and M2) relay control circuit shown. Include the starter's OL (overload relay contacts) programmed to turn on its associated pilot light should its OL contact trip open.

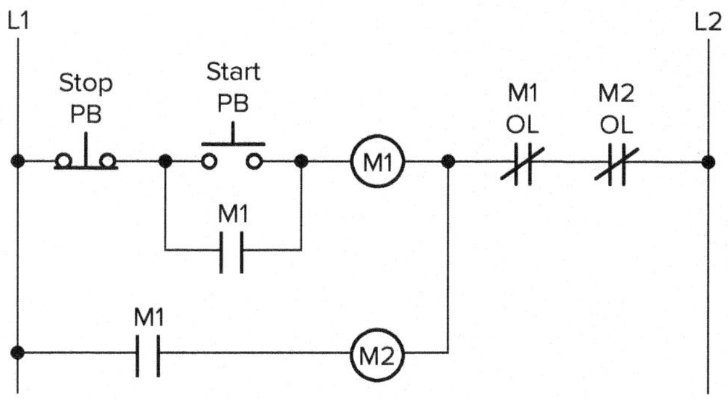

NAME ____________________ DATE__________

CHAPTER 3

Number Systems and Codes

3-1 The program shown is used to convert a binary number to its equivalent decimal value. Construct the program and test it by setting the input switches to various binary values and monitoring the resultant decimal displayed value.

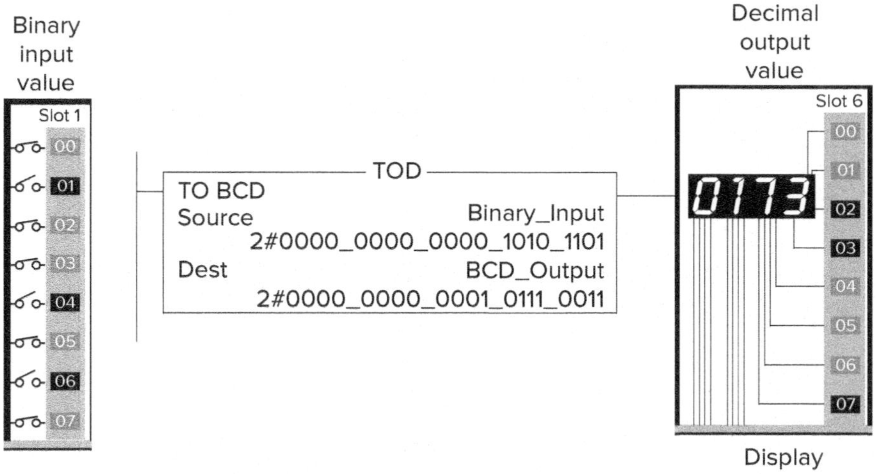

3-2 The program shown is used to view settings of a thumbwheel switch in an LED display monitor. Construct and test the program.

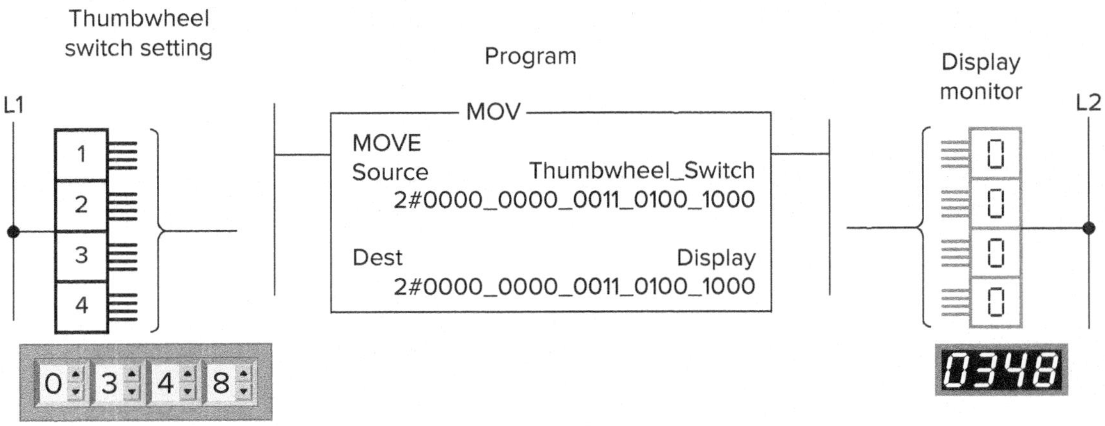

NAME ____ DATE________________

3-3 The program shown is used to monitor the accumulated time of a timer. Construct and test the program.

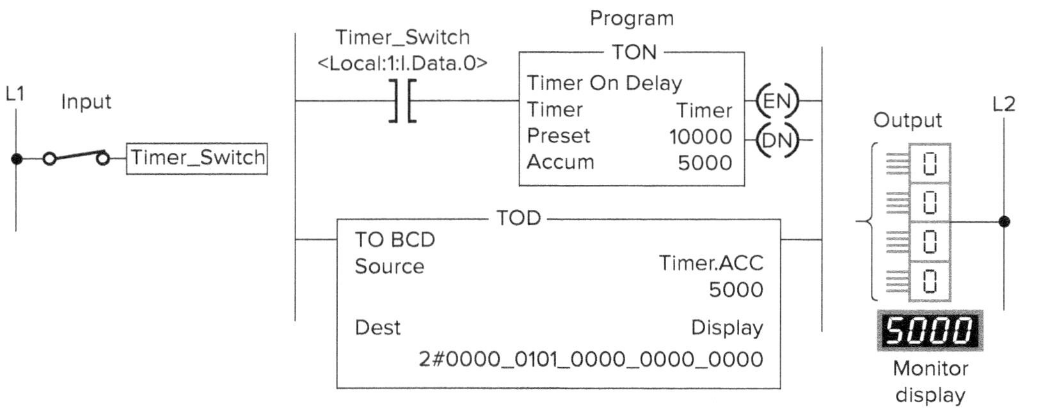

3-4 The program shown is used to compare the variable value of the thumbwheel switch (2) with a constant value of 5. Construct and test the program.

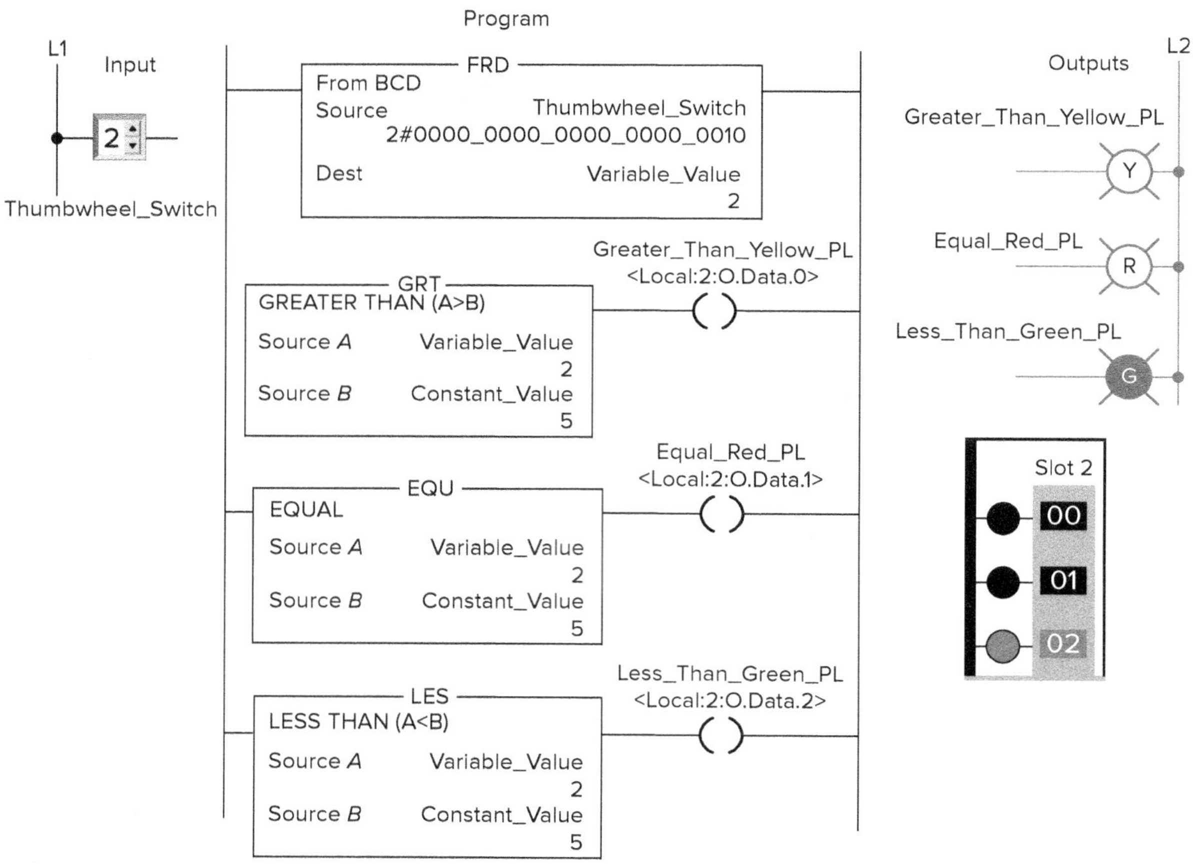

NAME ______________________ DATE ______________

CHAPTER 4

Fundamentals of Logic

4-1 The program shown is designed to illustrate **AND** logic. Construct and confirm the logic of the program.

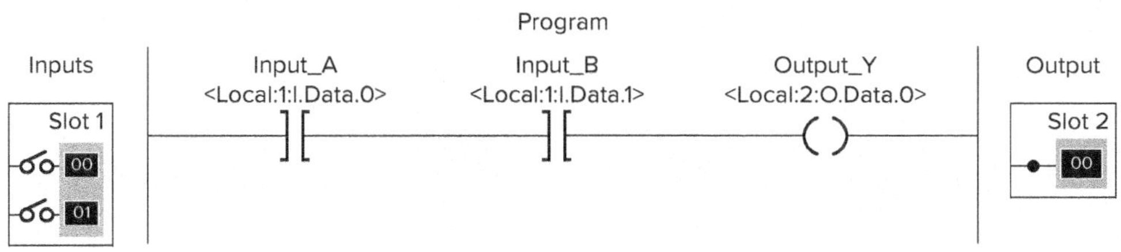

4-2 The program shown is designed to illustrate **OR** logic. Construct and confirm the logic of the program.

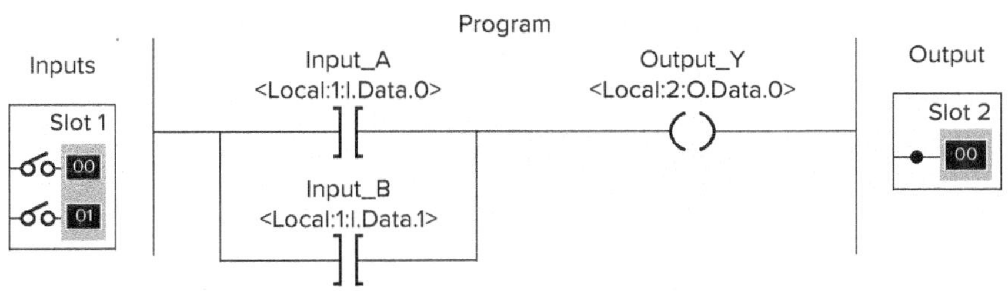

4-3 The program shown is designed to illustrate **NOT (inverter)** logic. Construct and confirm the logic of the program.

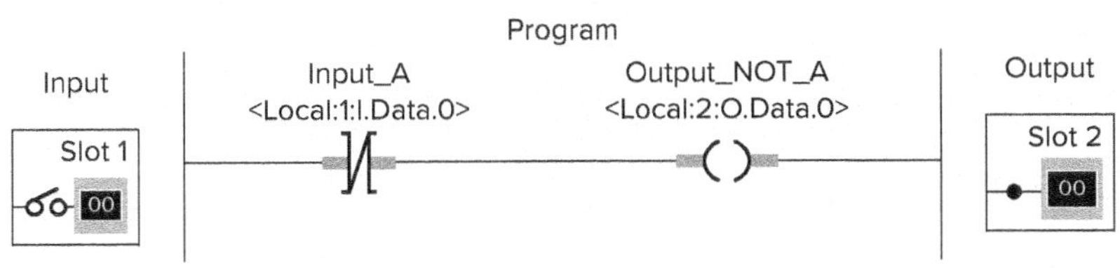

NAME ______________________________ DATE______________

4-4 The program shown is designed to illustrate **NAND** logic. Construct and confirm the logic of the program.

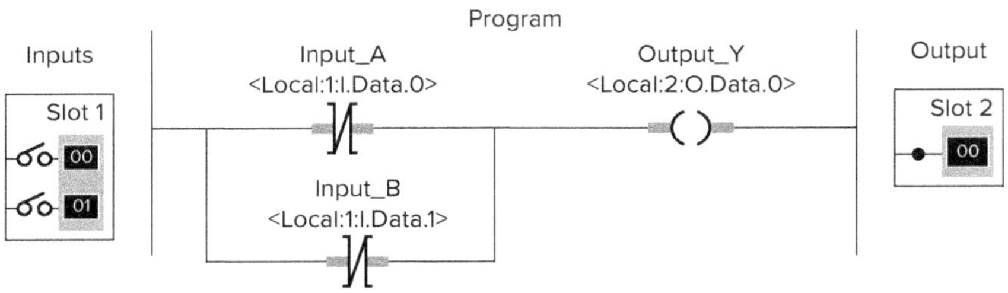

4-5 The program shown is designed to illustrate **NOR** logic. Construct and confirm the logic of the program.

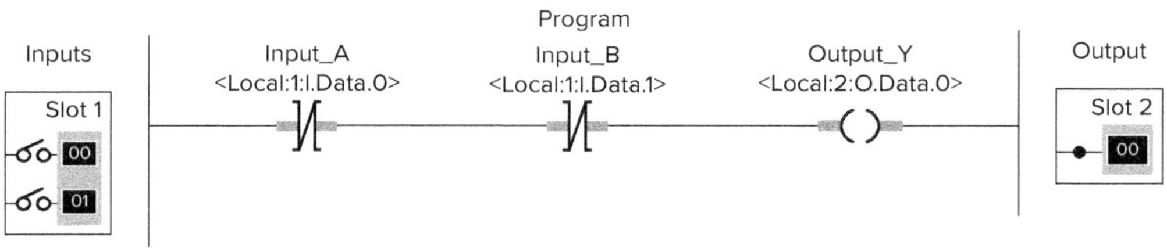

4-6 The program shown is designed to illustrate **XOR** logic. Construct and confirm the logic of the program.

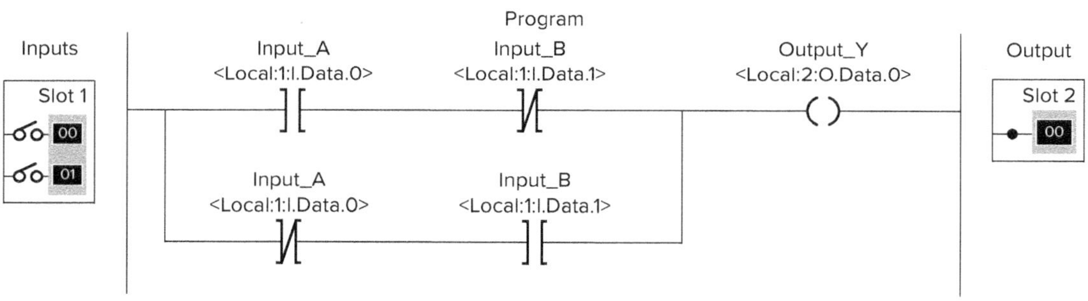

NAME ____________________ DATE ____________

4-7 Write and test a PLC program which will emulate the desired logic shown.

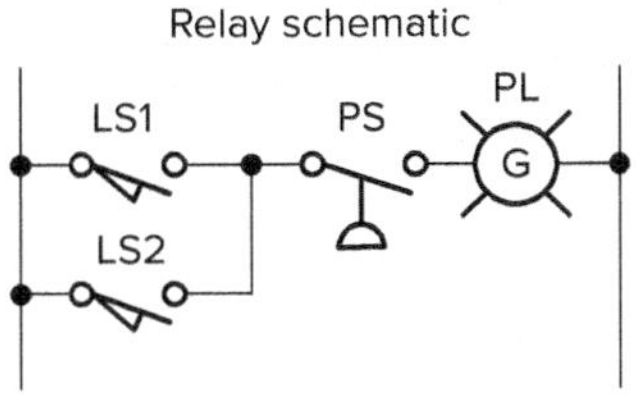

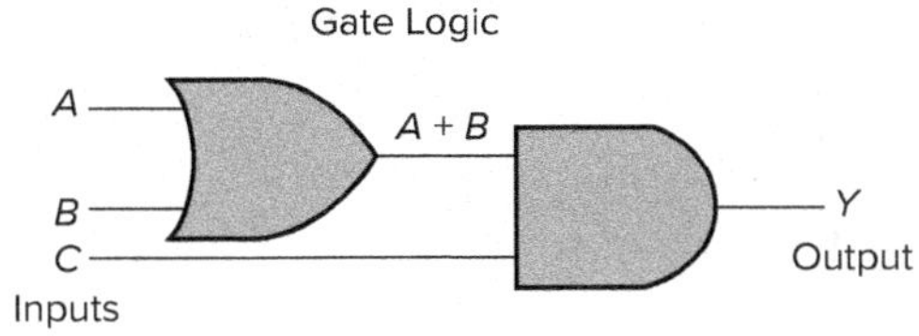

4-8 Write and test a PLC program which will emulate the desired logic shown.

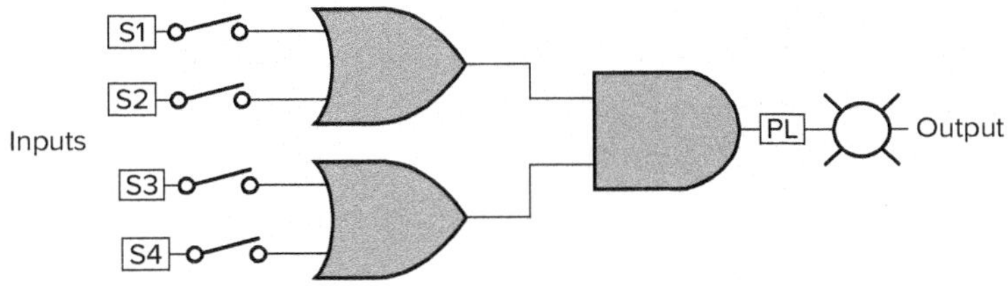

4-9 Write and test a PLC program which will emulate the desired logic shown.

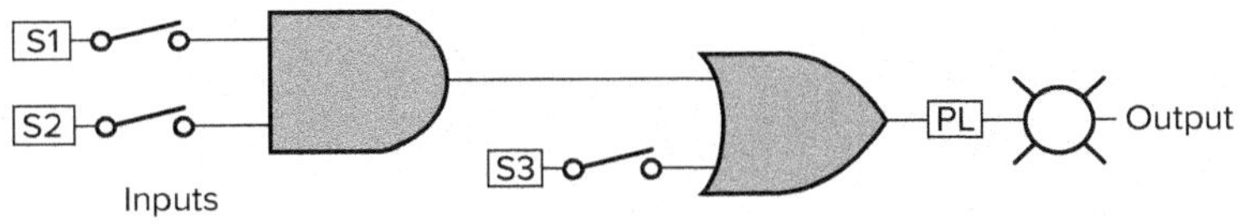

4-10 Write and test a PLC program which will emulate the desired logic shown.

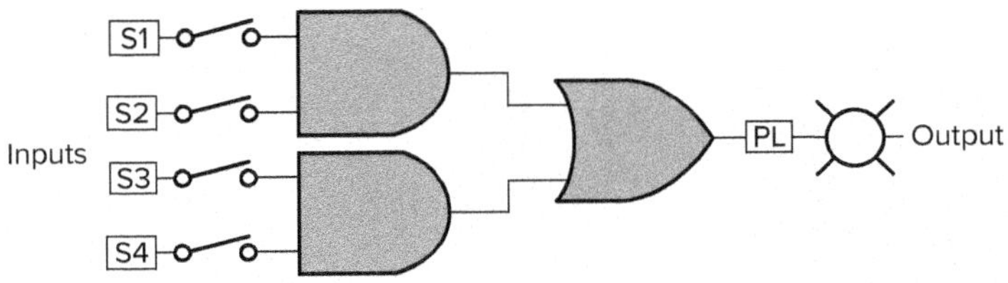

NAME ______________________________ DATE______________

4-11 Write and test a PLC program which will emulate the desired logic shown.

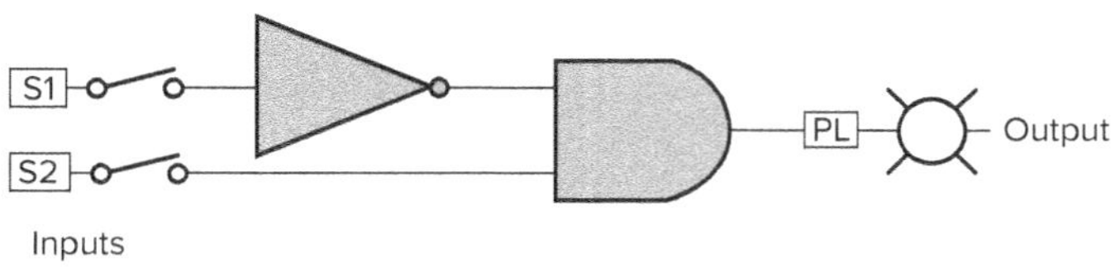

4-12 Write and test a PLC program which will emulate the desired logic shown.

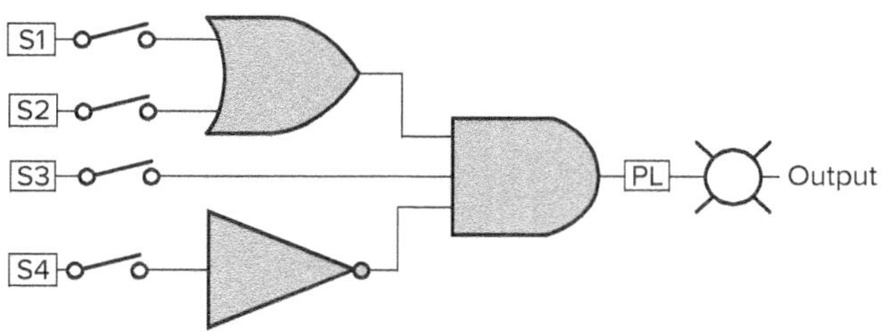

4-13 Construct the word-level AND instruction program shown. Execute your program by setting the ON/OFF (1 or 0) state of two input switches and viewing the resultant AND logic output.

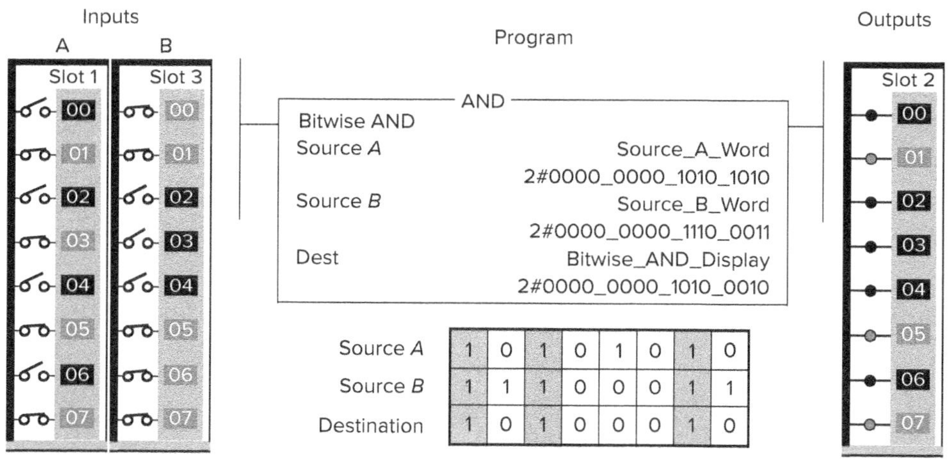

NAME ____________________ DATE ____________

4-14 Construct the word-level OR instruction program shown. Execute your program by setting the ON/OFF (1 or 0) state of two input switches and viewing the resultant OR logic output.

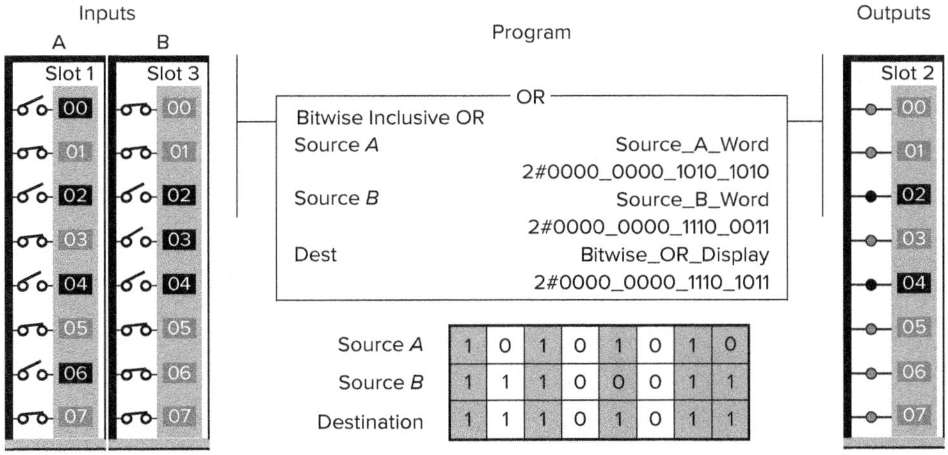

4-15 Construct the word-level XOR instruction program shown. Execute your program by setting the ON/OFF (1 or 0) state of two input switches and viewing the resultant XOR logic output.

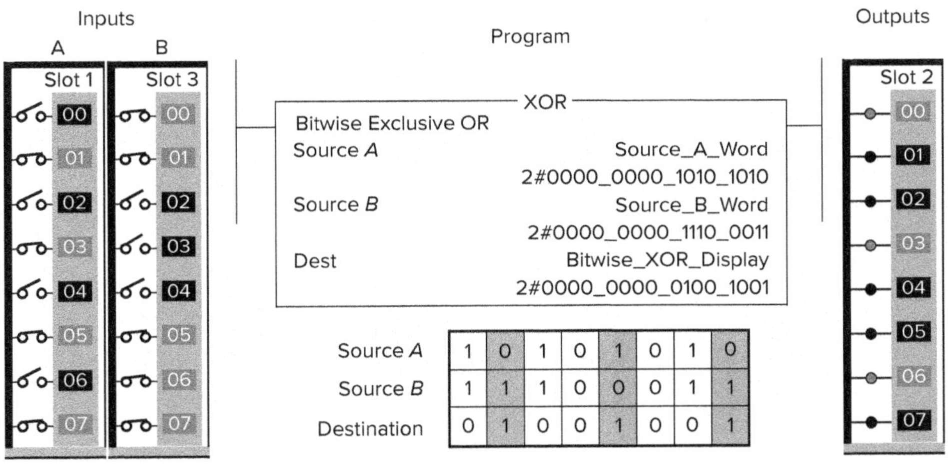

NAME ______________________________ DATE______________

CHAPTER 5

Basics of PLC Programming

5-1(a) Construct a PLC program which will emulate the tank Fill and Drain program shown. Run the program and Monitor the tags to verify its operation.

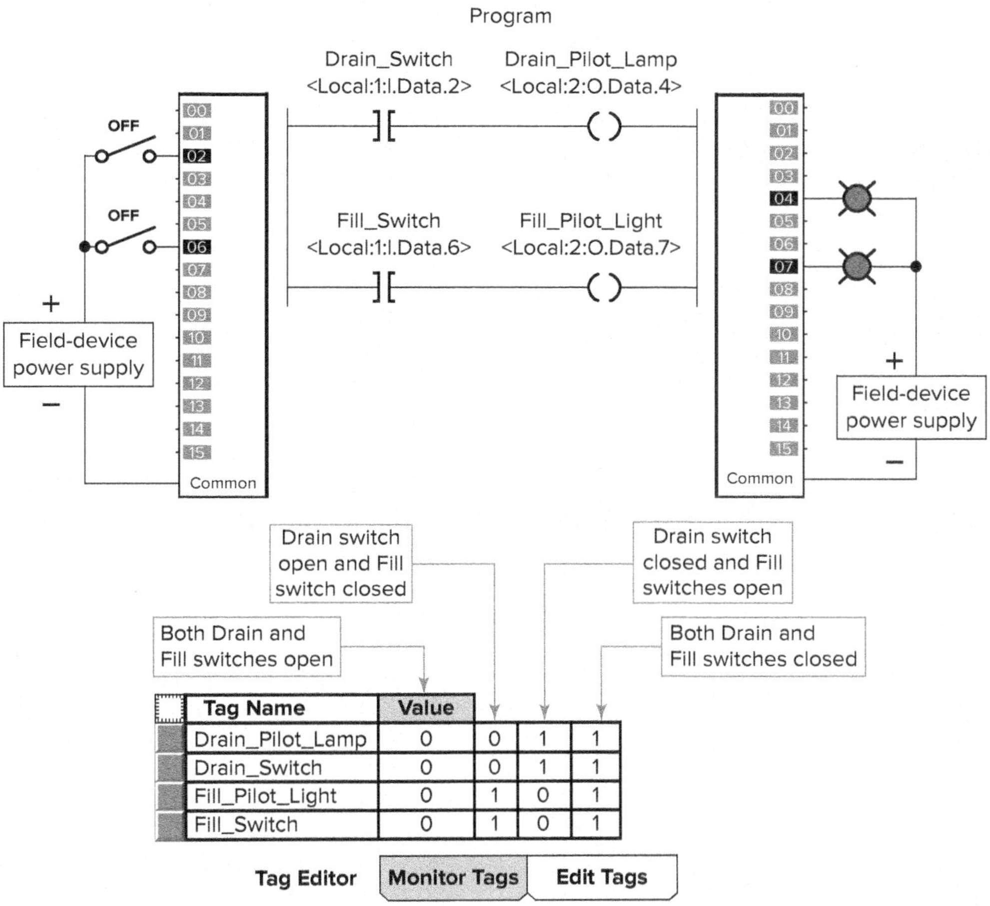

5-1(b) Modify the program so that the Fill and Drain process are prevented from being activated at the same time.

NAME ______________________________ DATE ______________

5-2 Write a PLC program for the desired logic shown. Confirm the logic of your program by monitoring the bit status of the tags relative to the truth table.

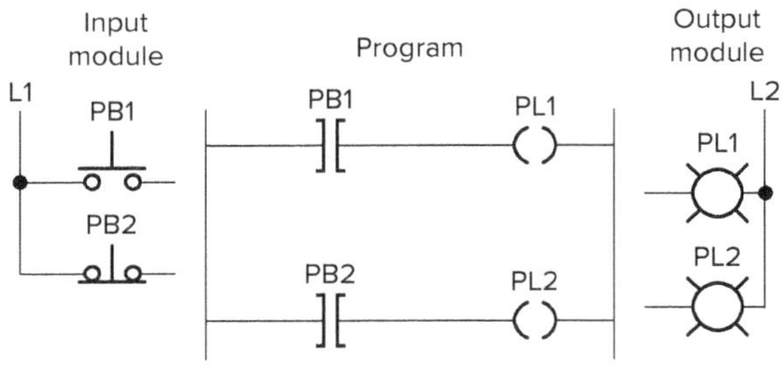

Truth Table

Input Device State	Input		Output	
	PB1	PB2	PL1	PL2
No PBs are pressed	0	1	0	1
PB1 is pressed	1	1	1	1
PB2 is pressed	0	0	0	0
Both PBs are pressed	1	0	1	0

5-3(a) Construct a PLC program exactly as shown. Operate your program and identify the error that exists.

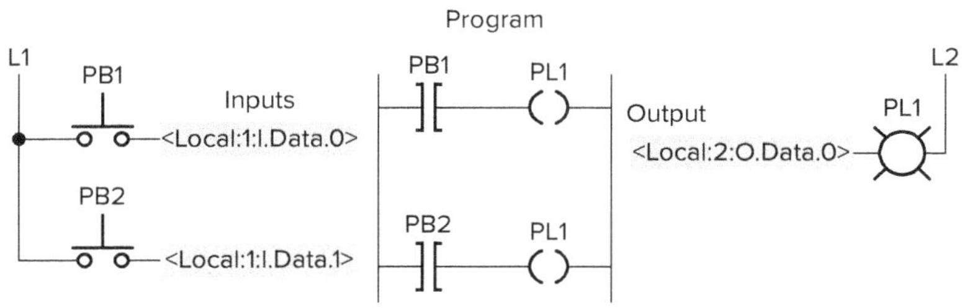

5-3(b) Modify the program to correct the error.

5-4 Write a PLC program for the desired logic shown. Confirm the logic of your program by monitoring the bit status of the tags relative to the truth table.

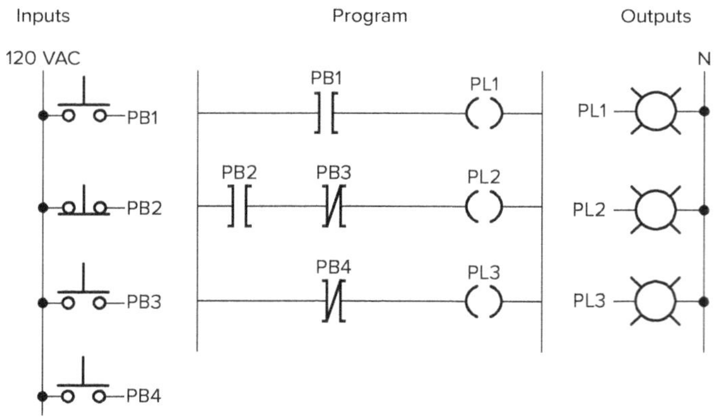

Truth Table

Input Device State	Input State				Output State		
	PB4	PB3	PB2	PB1	PL3	PL2	PL1
No PBs pressed	0	0	1	0	1	1	0
PB1 pressed	0	0	1	1	1	1	1
PB3 pressed	0	1	1	0	1	0	0
PB2,3,4 pressed	1	1	0	0	0	0	0

NAME ______________________________ DATE ______________

CHAPTER 6

Developing Fundamental PLC Wiring Diagrams and Ladder Logic Programs

6-1(a) Construct the PLC program for the automatic three-motor sequential starting circuit shown. Test your program for proper operation.

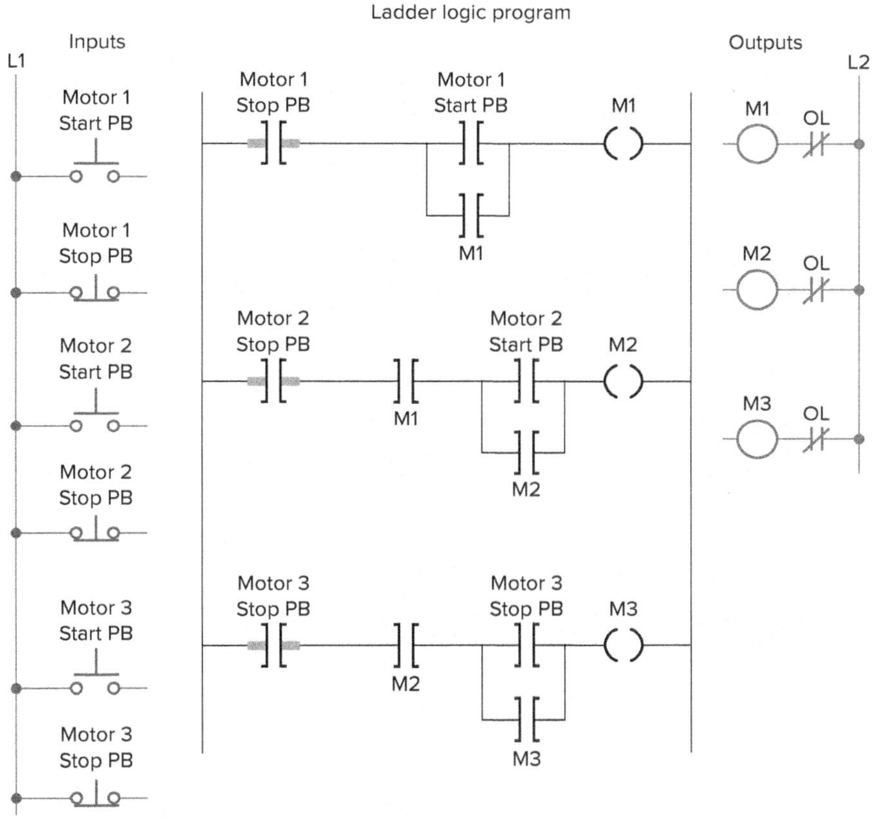

NAME ______________________ DATE______________

6-1(b) Modify your program for a Motor 1, Motor 3, Motor 2, operating sequence.

6-2(a) Construct the PLC pushbutton interlocking program shown. Test your program for proper operation.

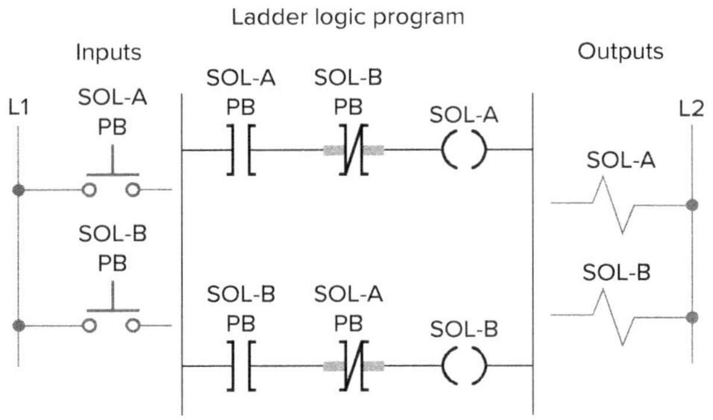

6-2(b) Expand your program to include a PL which will be energized whenever both pushbuttons are pressed closed at the same time.

6-3(a) Construct the Forward and Reverse motor control program shown. Test your program for proper operation.

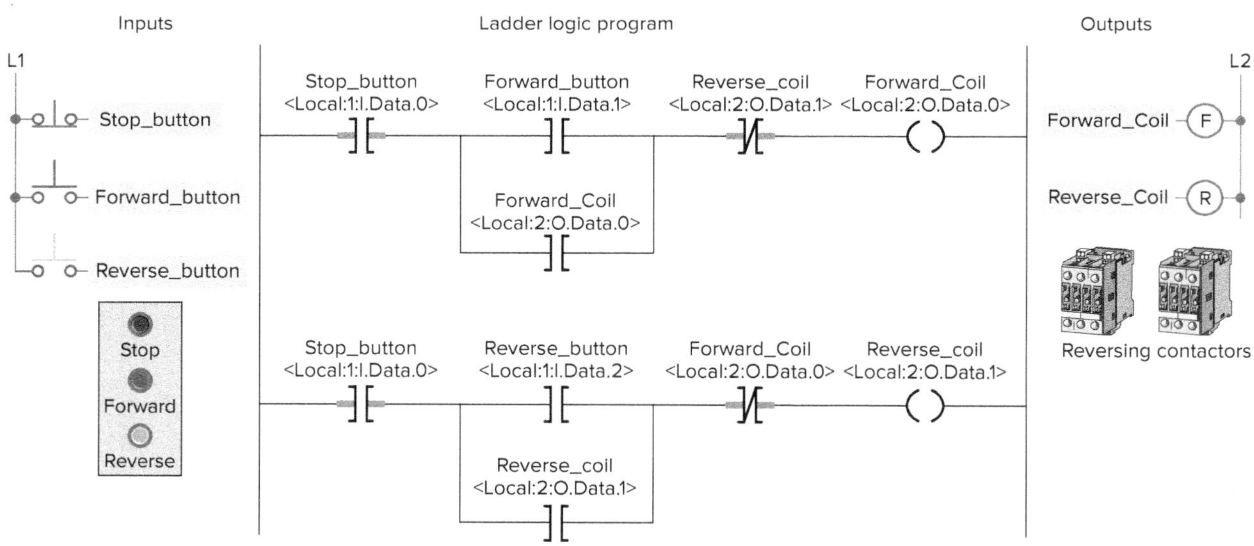

NAME ________________________________ DATE______________

6-3(b) Modify your program to include pushbutton interlocking as illustrated in the relay control schematic shown.

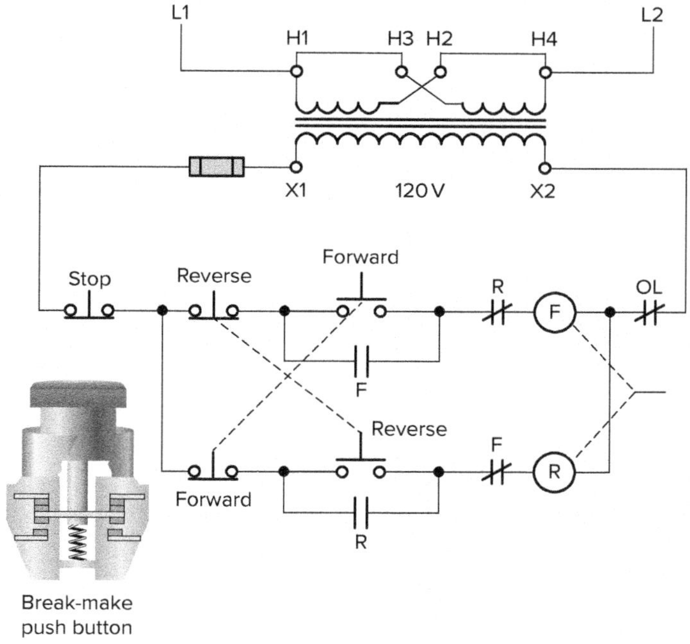

6-4(a) Construct the jog-motor control program shown. Test your program for proper operation.

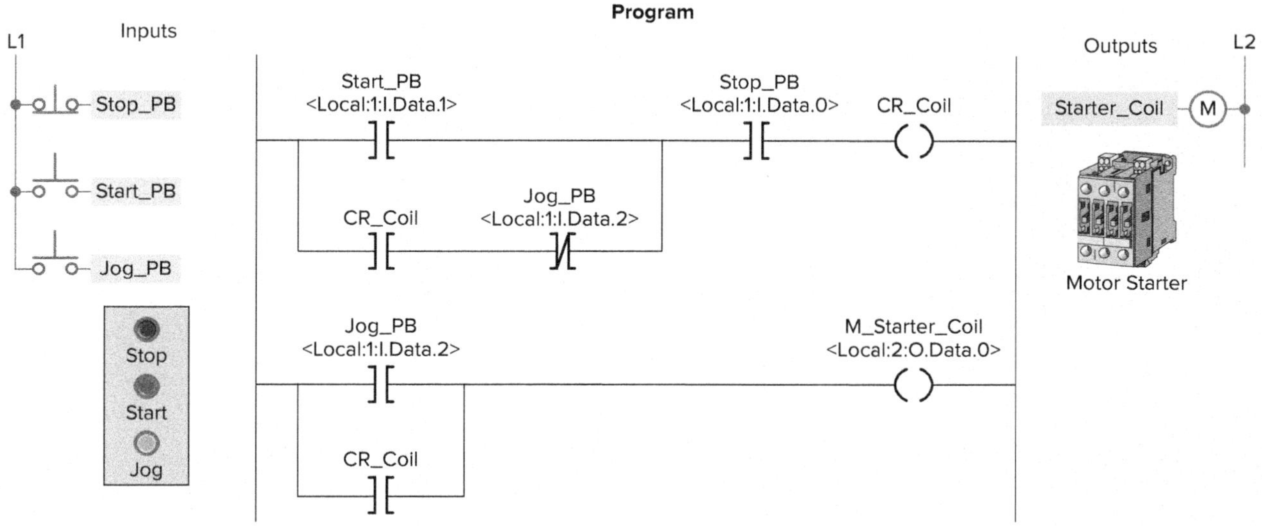

NAME ______________________ DATE______________

6-4(b) Modify your program to include a PL which is energized anytime the program is operated in the jog mode.

6-5 Construct the simple Latch and Unlatch program shown. Test program for proper operation by monitoring its status bit.

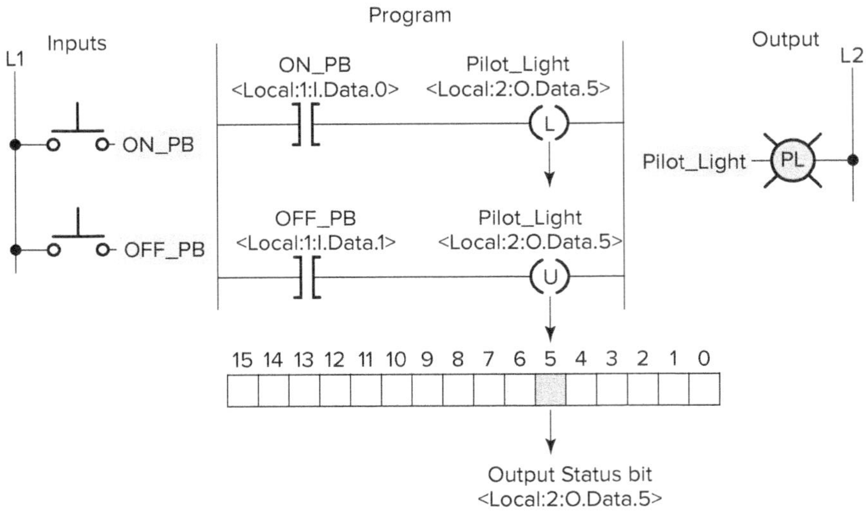

6-6 Construct the program shown to control water level in a storage tank. Test your program for proper operation.

NAME ______________________________ DATE ______________

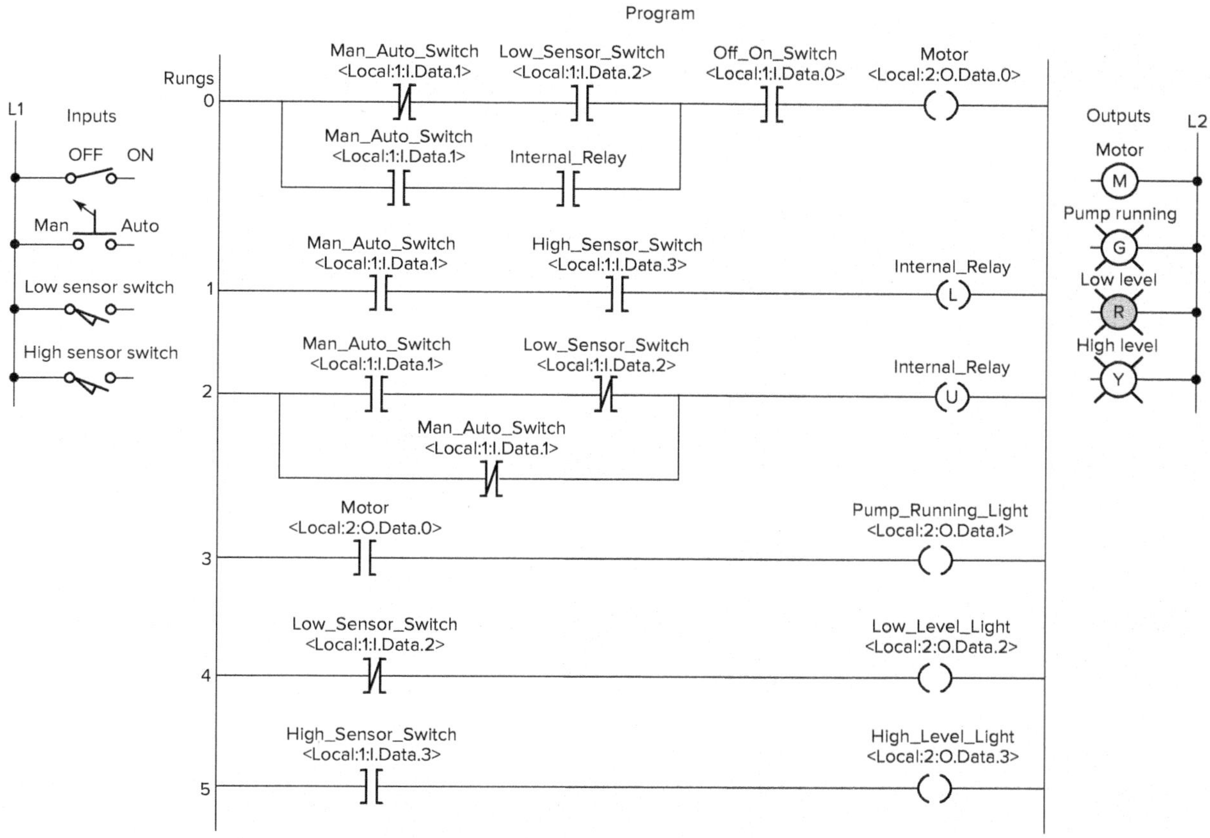

6-7 Construct the push-to-start/push-to-stop pushbutton PLC program shown. Test your program for proper operation as outlined in the text.

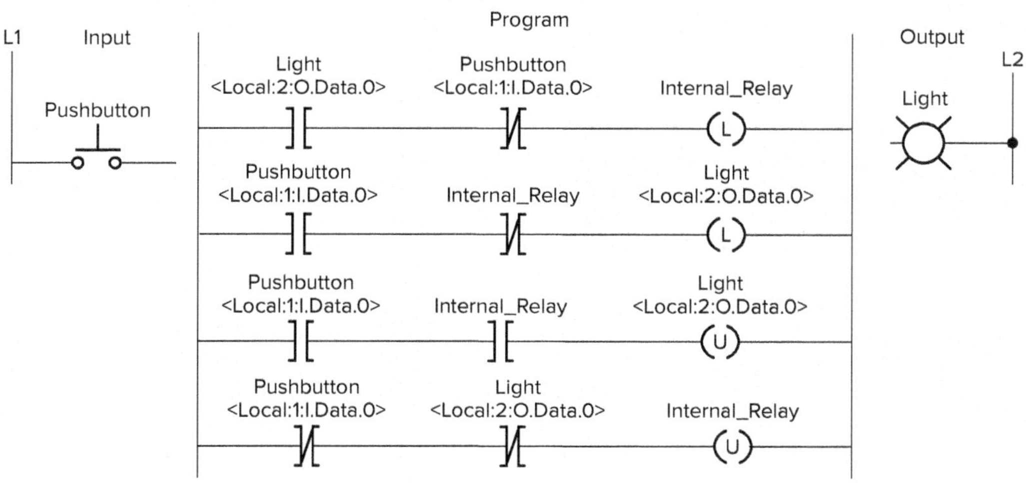

NAME ______________________________ DATE ________________

6-8 Construct the dual compressor operation program shown. Test your program for proper operation as outlined in the text.

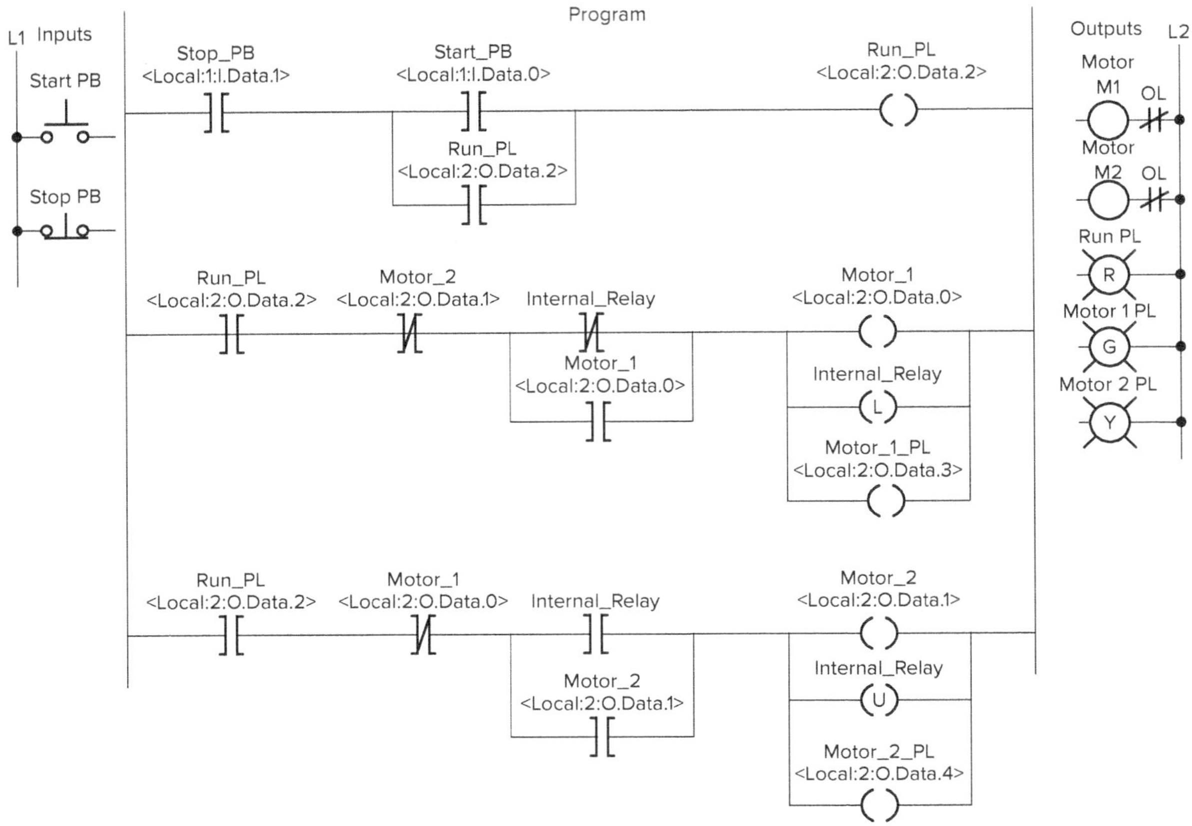

6-9 Construct the car wash program shown. Test your program for proper operation.

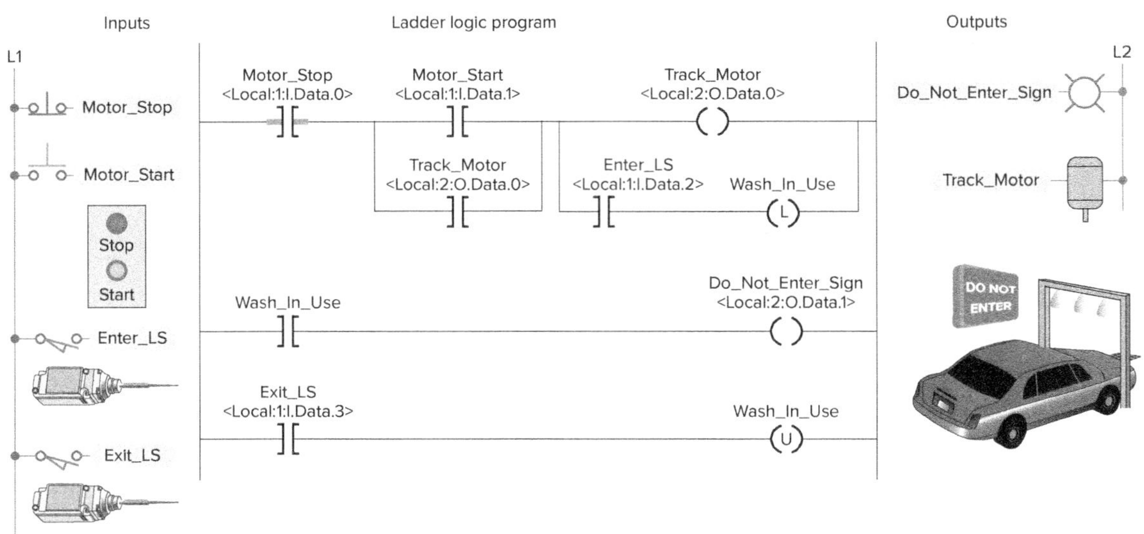

NAME ______________________________ DATE______________

6-10 Construct the PLC sequential process program shown. Test your program for proper operation.

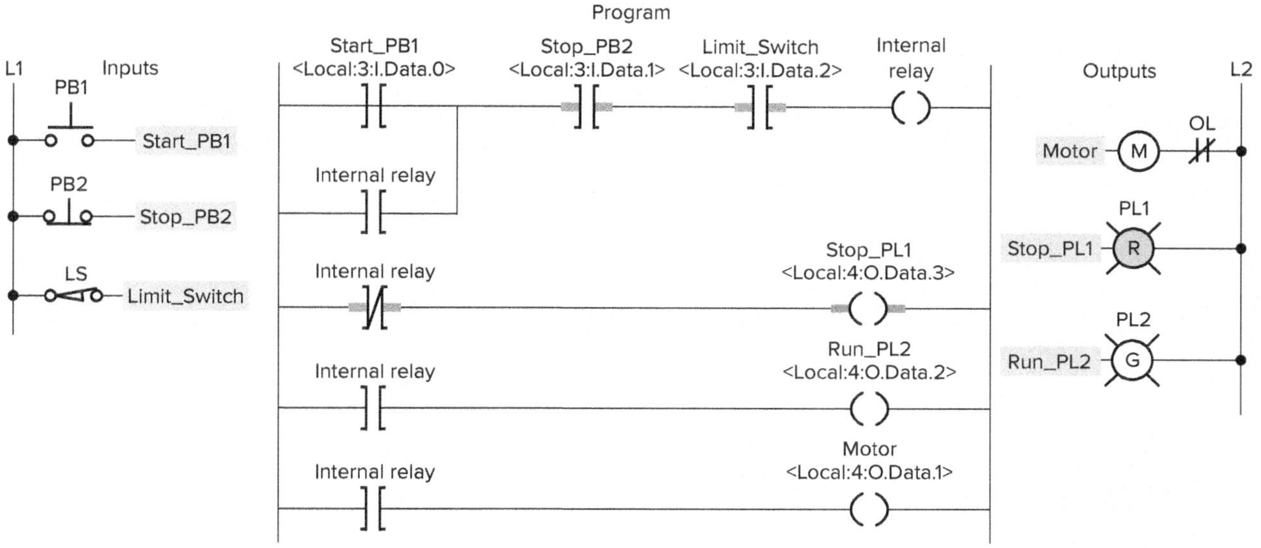

6-11 Construct the motorized overhead garage door PLC program shown. Test your program for proper operation.

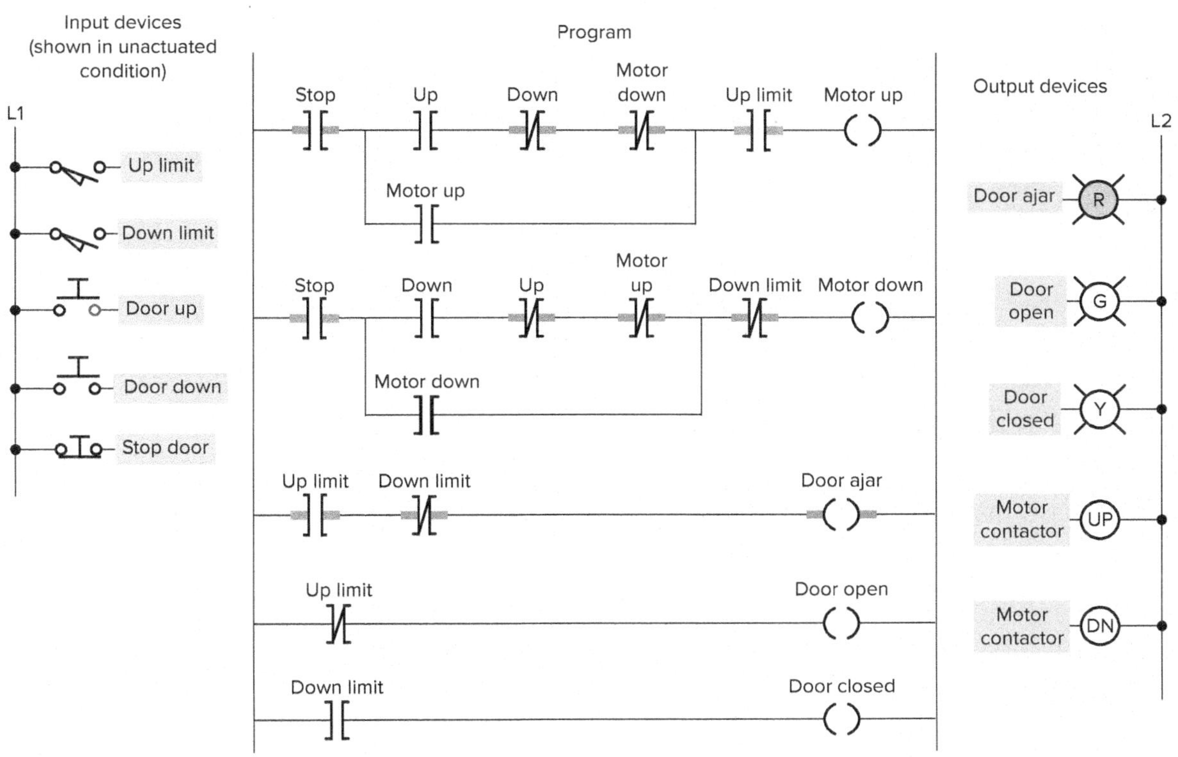

NAME ________________________________ DATE _______________

6-12 Construct the continuous filling operation PLC program shown. Test your program for proper operation.

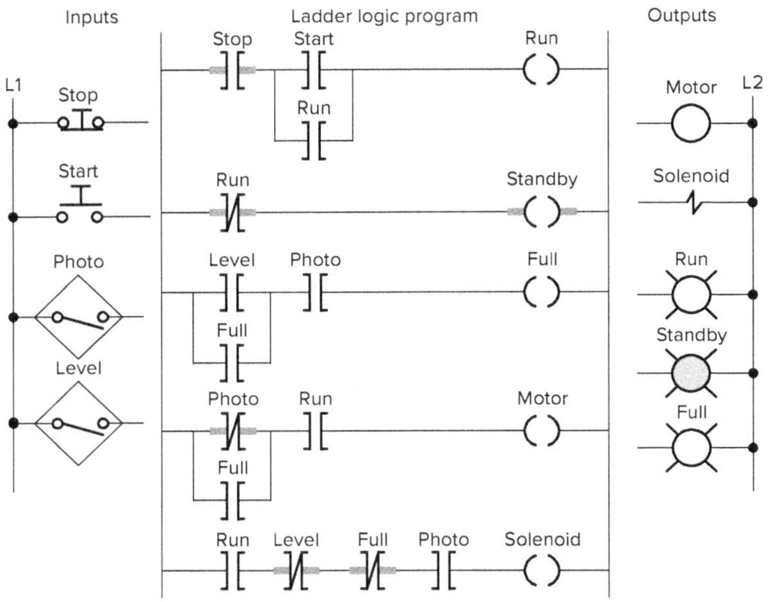

6-13 Write a PLC program that will perform the following sequential switching tasks using 4 single pole switches for inputs and 3 pilot lights for outputs:

- When switch S1 is closed, lights L1, L2, and L3 turn on.
- When switch S1 and S2 are closed, lights L1 and L2 turn off while light L3 remains on.
- When switches S1, S2, and S3 are closed, light L1 remains on, L2 is turned off and L3 is turned on.
- Any time switch S4 is closed it turns off any of the lights that happen to be on.

6-14 Write a PLC program that will perform the following Latch/Unlatch operation using 3 single pole switches (S1, S2, and S3) for inputs and single pilot light (PL) for the output:

- PL to be latched on when S1 is momentarily closed and unlatched when either S2 or S3 switch is momentarily closed.
- Do not allow the unlatch rung to go true when the latch rung is true.
- Do not allow the latch instruction to go true when the unlatch rung is true.

NAME ________________________________ DATE______________

6-15 Write a PLC program that will perform the following monitoring of a process for malfunctions using 4 single pole normally open switches (S1, S2, S3, S4) for inputs and 3 color coded pilot lights (Green, Yellow, Red) for the outputs:

- If any one sensor input is closed nothing happens.
- When any two sensor inputs are closed the Green PL is switched on.
- When any three sensor inputs are closed the Yellow PL is switched on.
- When all 4 sensor inputs are closed the Red PL is switched on.

6-16 Write a PLC program that will implement the accompanying sequential relay control circuit.

- The operator is required to actuate two pushbuttons, one at a time, in the proper sequence to initiate the process.
- The correct sequence is PB1 followed by PB2.
- Once the two buttons have been pressed in the correct sequence, the Correct Sequence light will come on and stay on until the reset button is pressed.
- If the wrong sequence occurs, the error light comes on and stays on until the reset button is pressed.
- Once an error has occurred, no sequence of operation of the pushbuttons can start the process.
- Utilize internal relays within the program.

NAME ____ DATE_________________

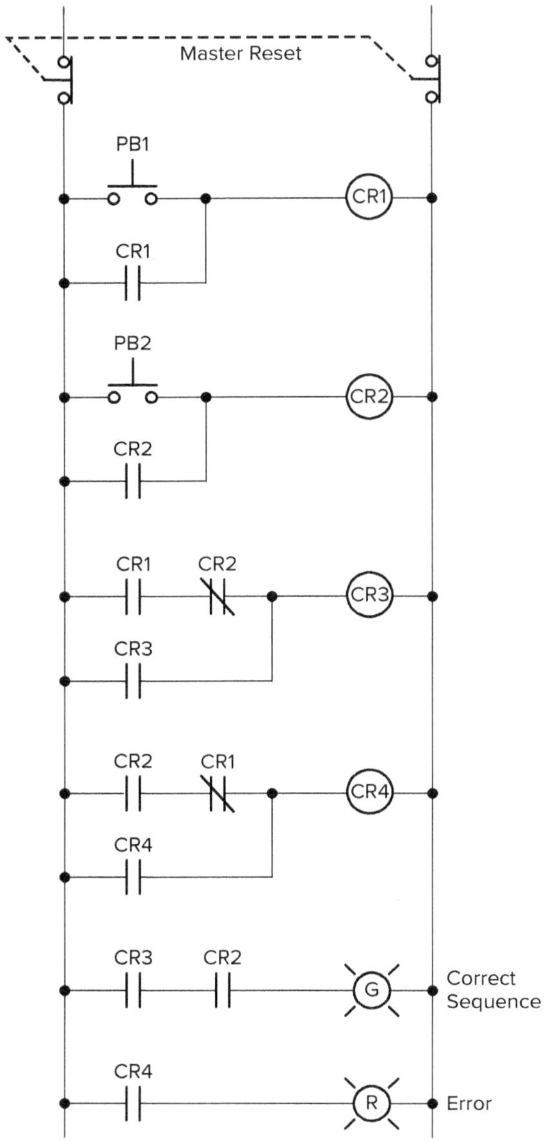
Master Reset
PB1
CR1
CR1
PB2
CR2
CR2
CR1
CR2
CR3
CR3
CR2
CR1
CR4
CR4
CR3
CR2
G
Correct
Sequence
CR4
R
Error

NAME ________________________________ DATE ______________

CHAPTER 7

Programming Timers

7-1 Construct a PLC program for the accompanying hardwired on-delay timer circuit shown. The delay time is to be 10 seconds and the timer contact used is a normally open timed closed **(NOTC)** type. Test your program for proper operation.

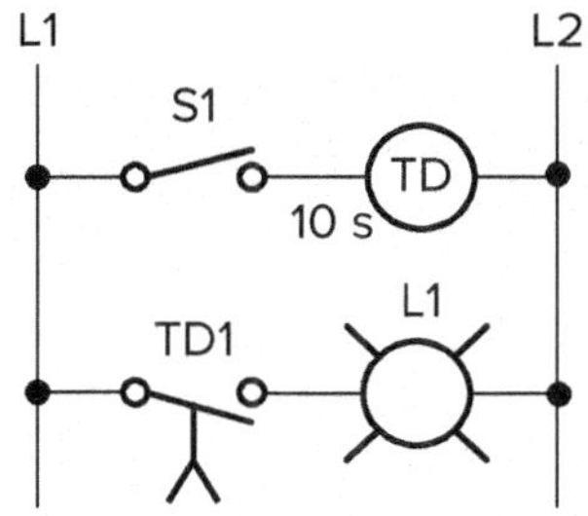

7-2 Construct a PLC program for the accompanying hardwired on-delay timer circuit shown. The delay time is to be 10 seconds and the timer contact used is a normally closed timed open **(NCTO)** type. Test your program for proper operation.

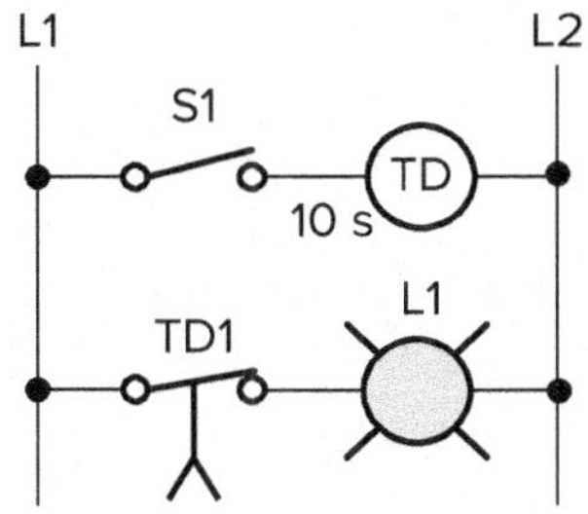

7-3 Construct a PLC program for the accompanying hardwired off-delay timer circuit shown. The delay time is to be 10 seconds and the timer contact used is a normally open timed open contact **(NOTO)** type. Test your program for proper operation.

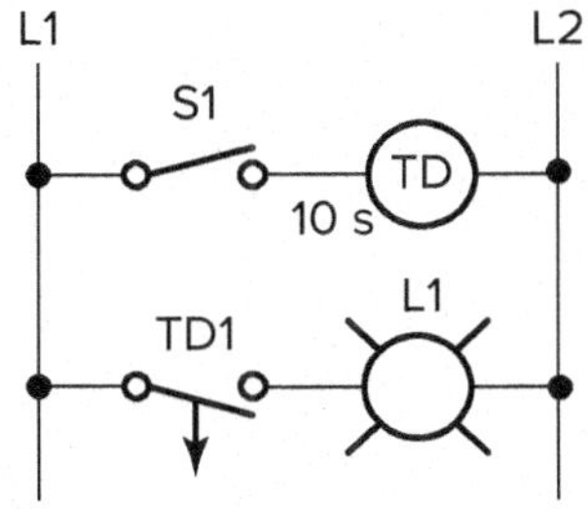

7-4 Construct a PLC program for the accompanying hardwired off-delay timer circuit shown. The delay time is to be 10 seconds and the timer contact used is a normally closed timed closed **(NCTC)** type. Test your program for proper operation.

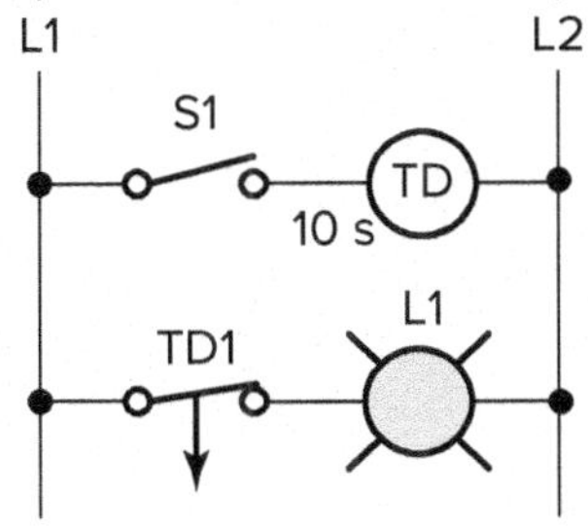

NAME ________________________________ DATE______________

7-5 Construct the PLC on-delay timer instruction program shown. Test your program for proper operation.

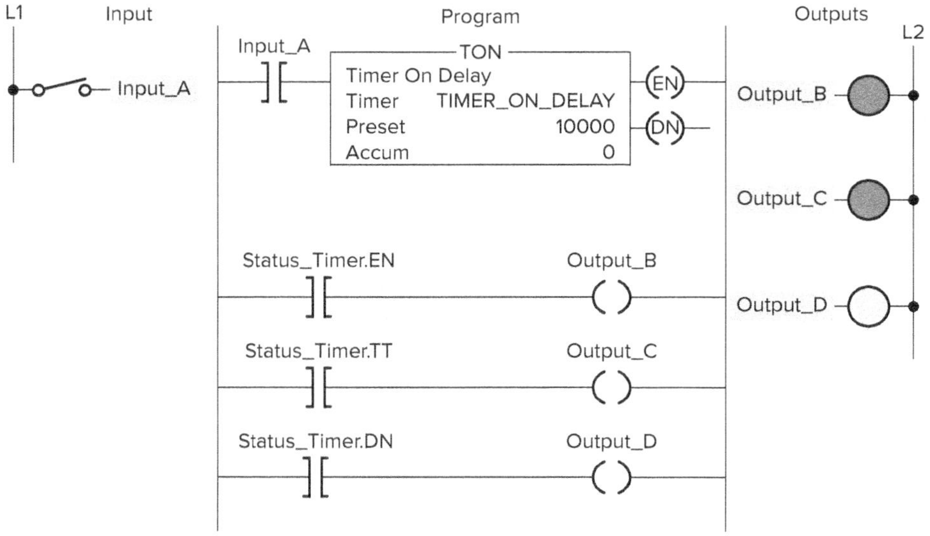

7-6(a) Construct the TON conveyor warning signal program shown. Test your program for proper operation.

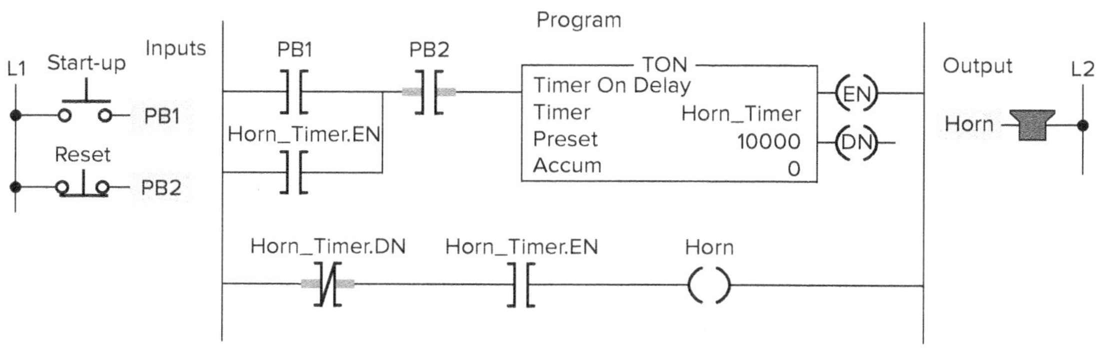

7-6(b) Add a pilot light to indicate the conveyor is running.

7-7 Construct a program for the automatic sequential control process shown. Test your program for proper operation.

NAME ______________________ DATE ______________

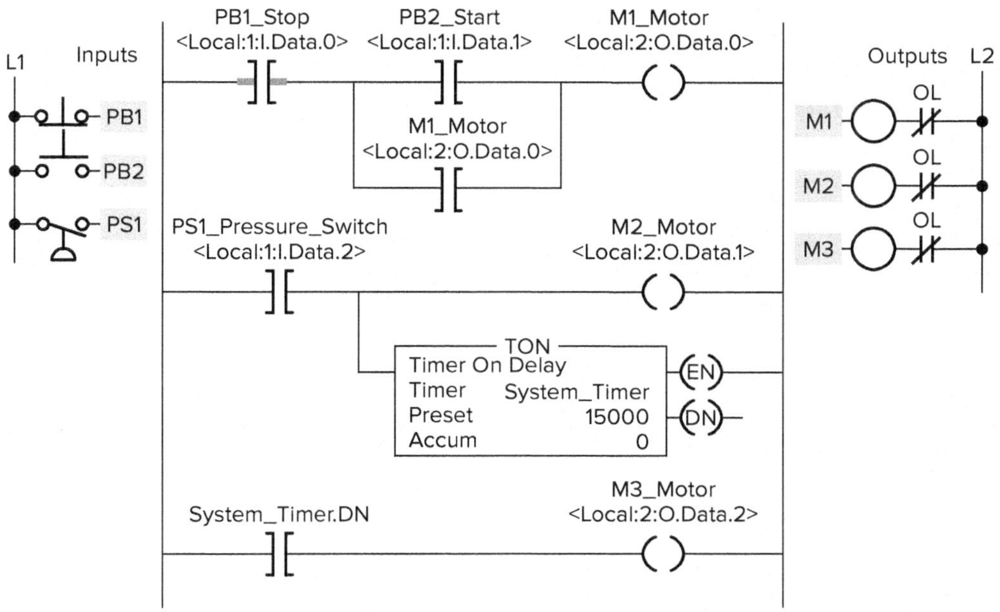

7-8 Construct a program for the TON timer used in the operation of a diverter gate solenoid shown. Test your program for proper operation.

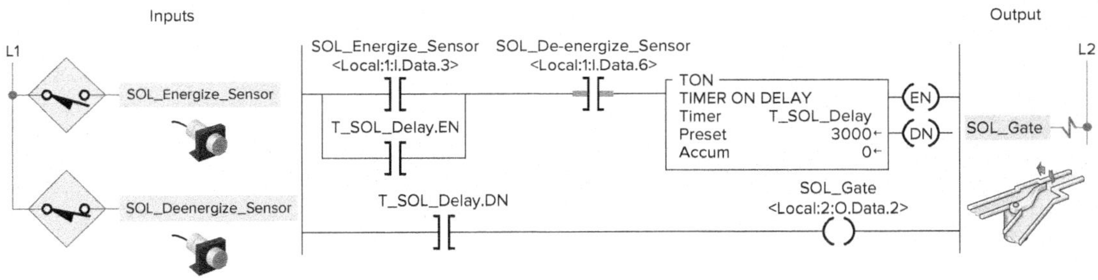

7-9 Construct a program for the pilot light TON timer program shown. Test your program for proper operation.

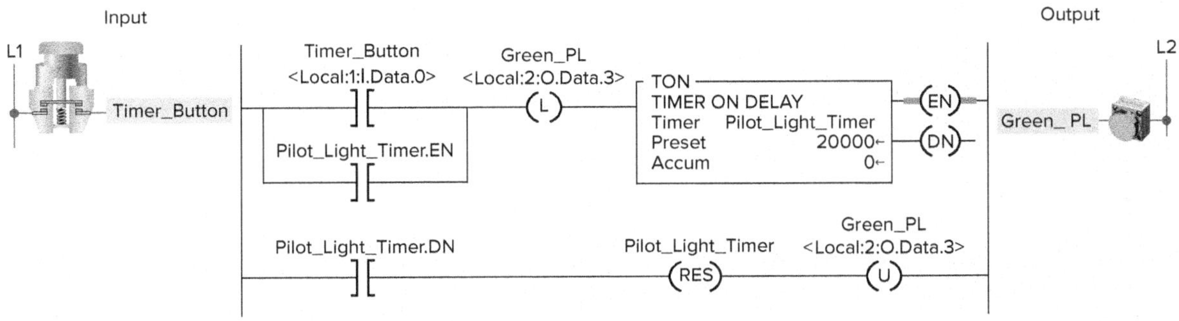

NAME ____________________ DATE__________

7-10 Construct a program for the pilot light TOF timer program shown. Test your program for proper operation.

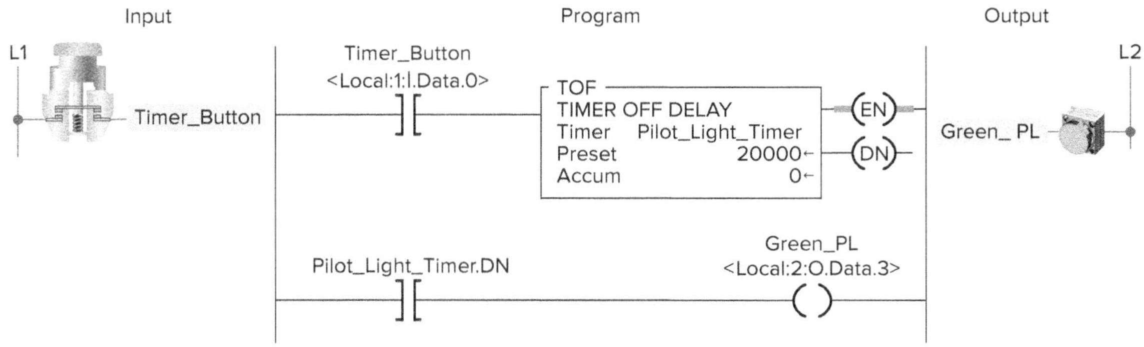

7-11 Construct a program for the timed motor switching program shown. Test your program for proper operation.

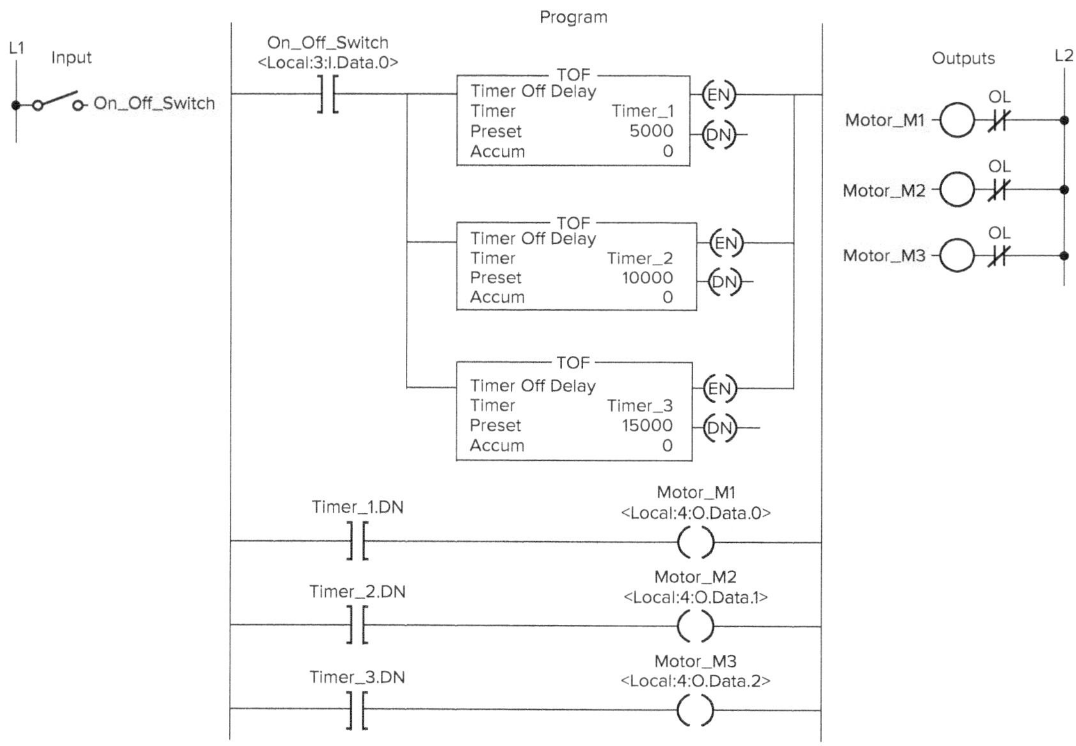

NAME ____________________ DATE ____________

7-12 Construct a program for the off-delay timer with instantaneous contacts program shown. Test your program for proper operation.

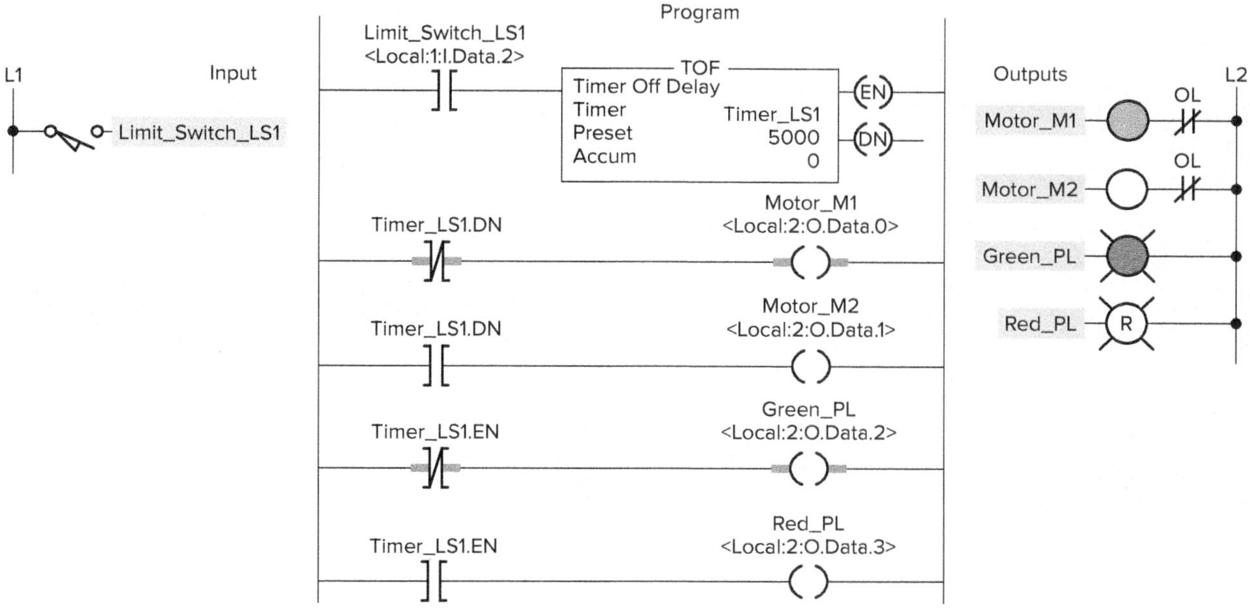

7-13 Construct a program for the fluid pumping process program shown. Test your program for proper operation.

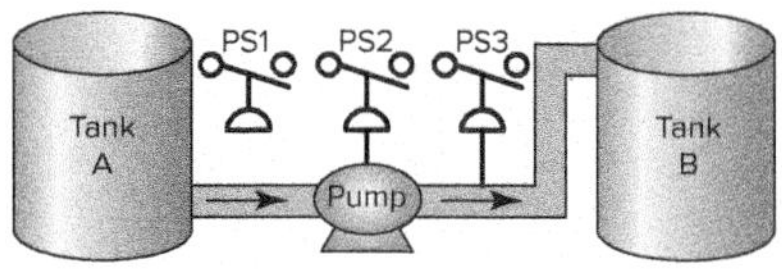

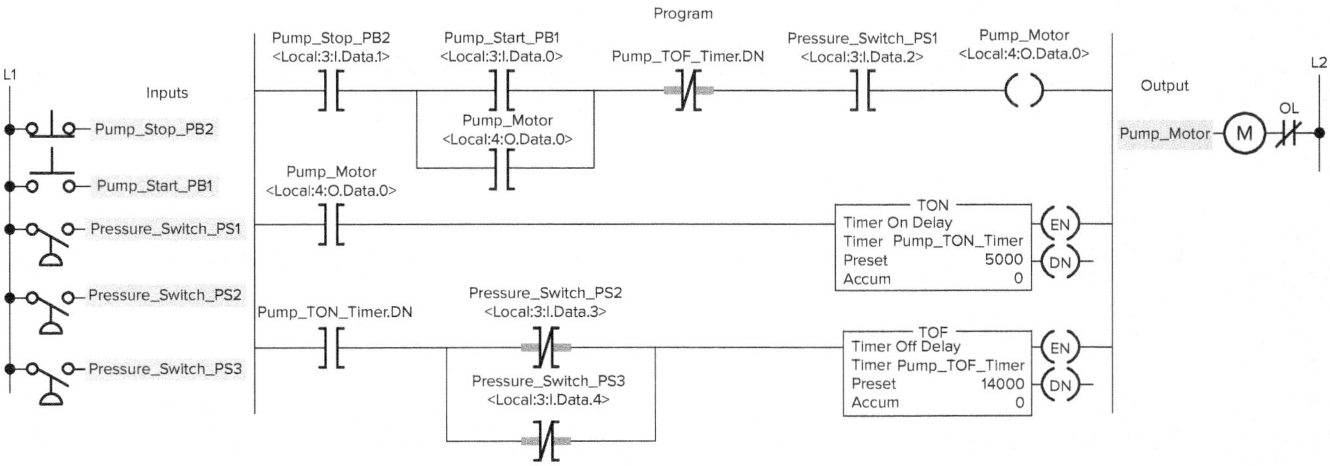

NAME ____________________ DATE ______________

7-14 Construct a program for the heating oven process program shown. Test your program for proper operation.

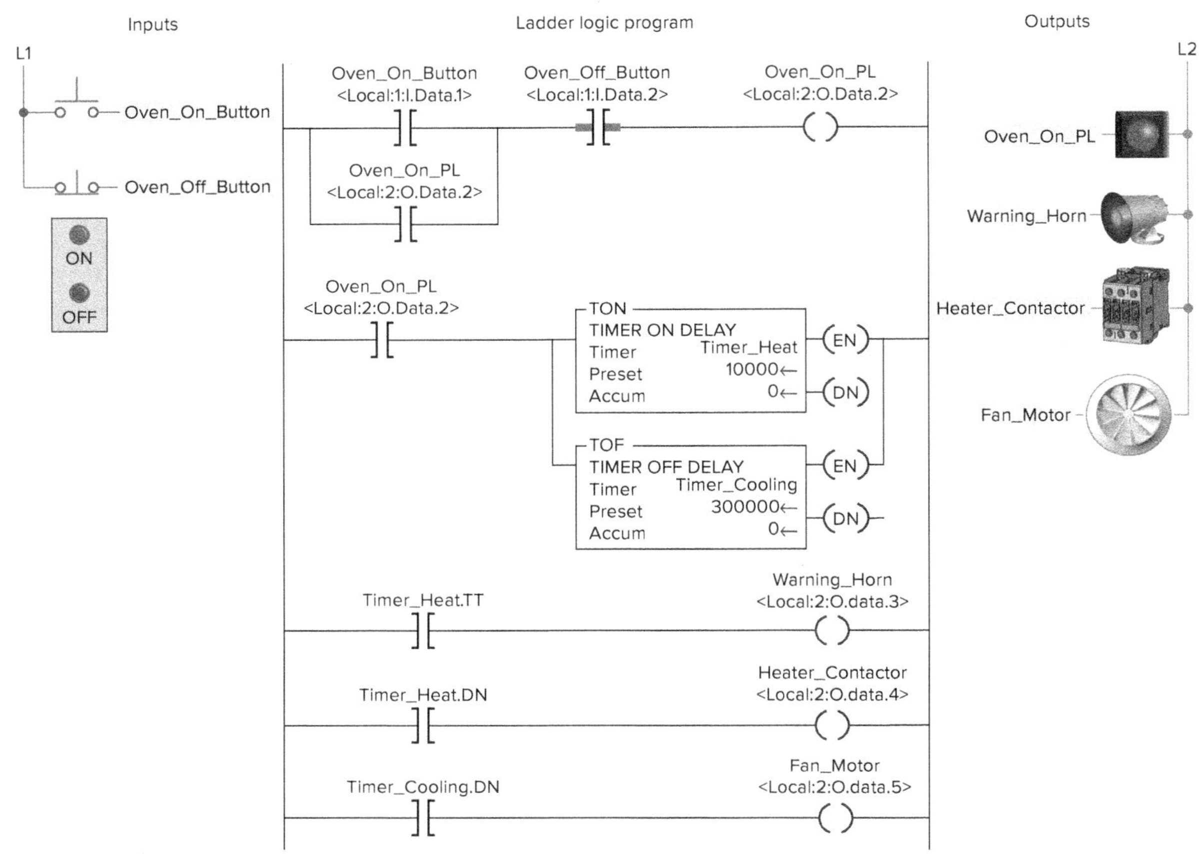

7-15 Construct a program for the retentive on-delay timer operation shown. Test your program for proper operation.

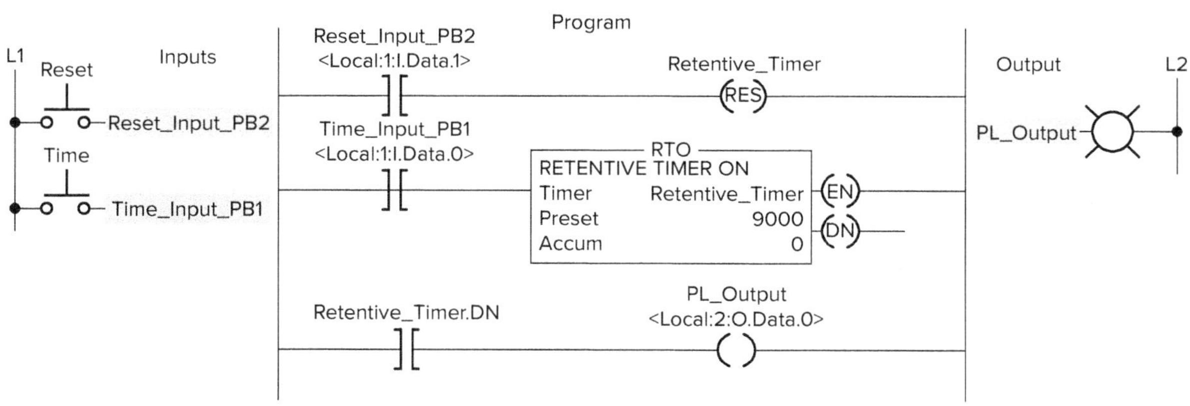

NAME ________________________ DATE ____________

7-16 Construct a program for the pressure alarm program shown. Test your program for proper operation.

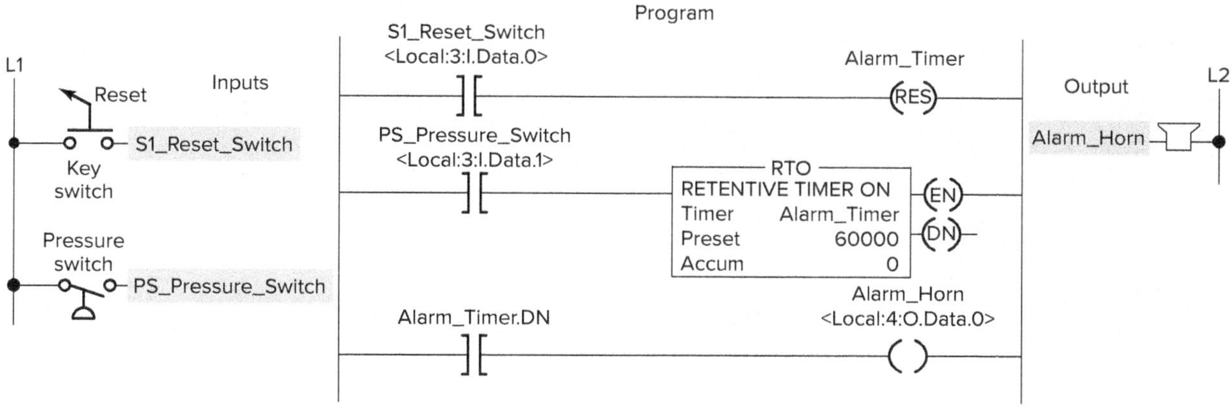

7-17 Construct a program for the bearing lubrication program shown. Test your program for proper operation.

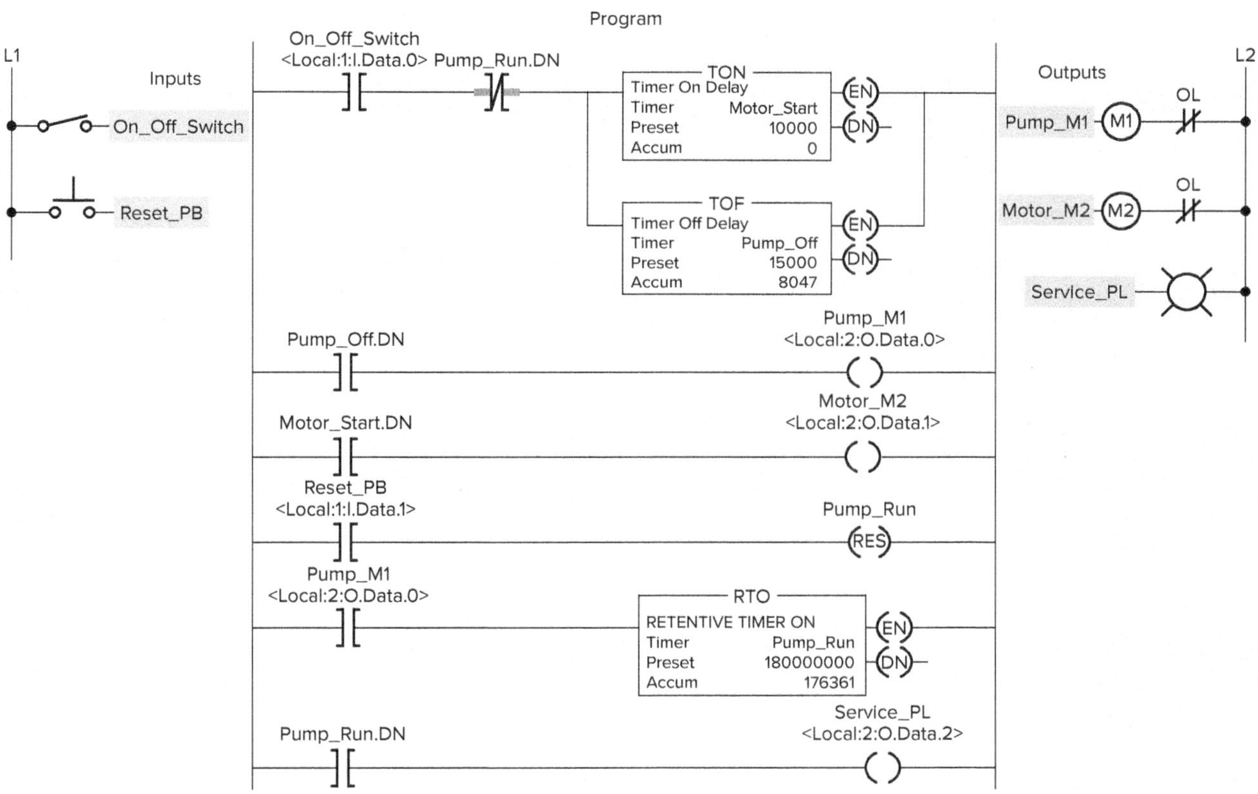

NAME ______________________________ DATE ______________

7-18 Construct a program for the time-delayed motor-starting program shown. Test your program for proper operation.

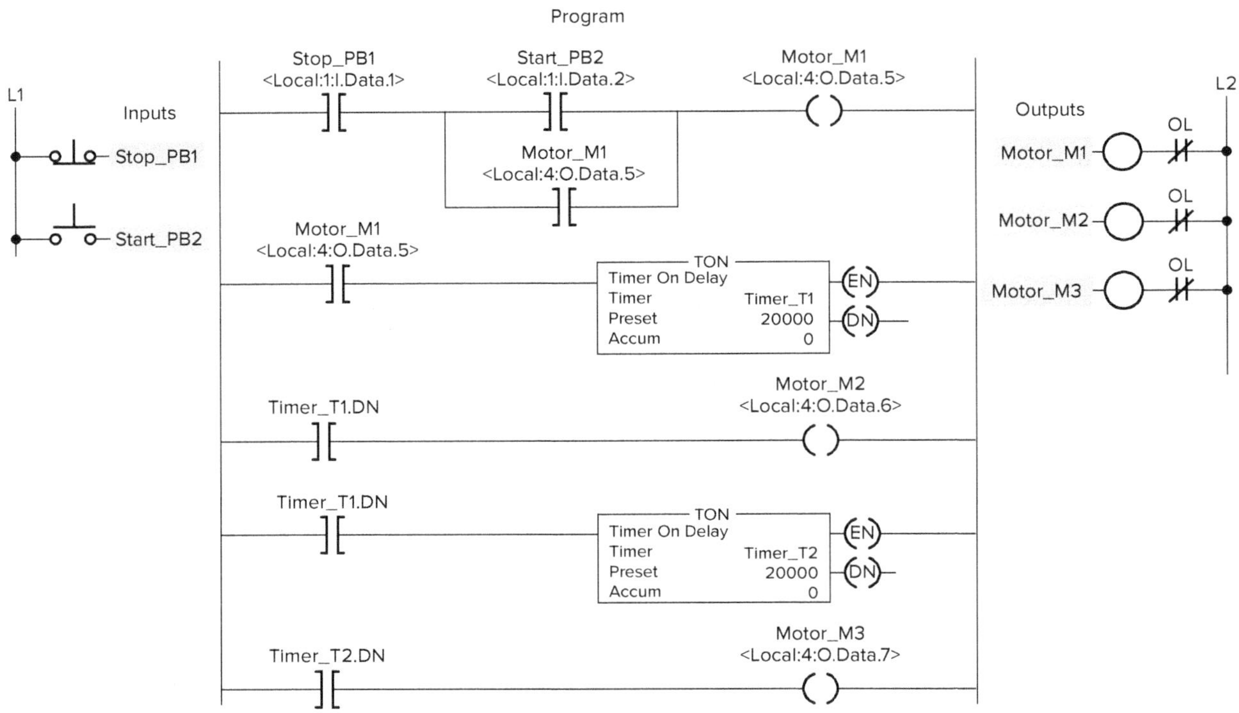

7-19 Construct a program for the annunciator flasher program shown. Test your program for proper operation.

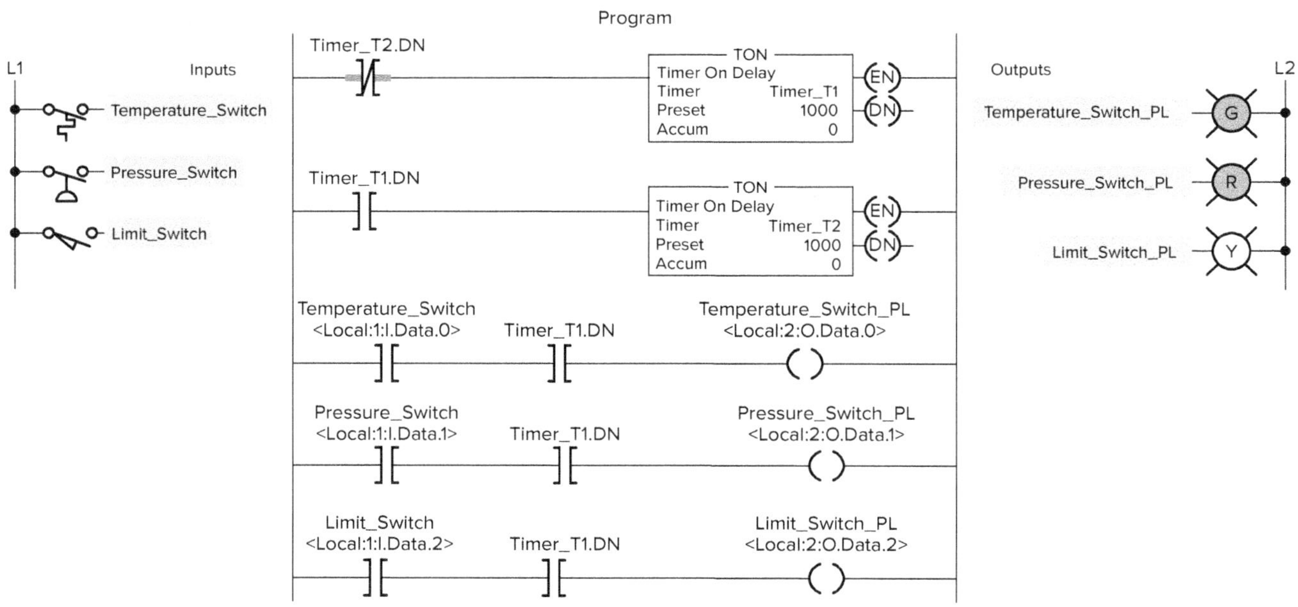

NAME ______________________________ DATE______________

7-20 Construct a program for the two-direction traffic light control program shown. Test your program for proper operation.

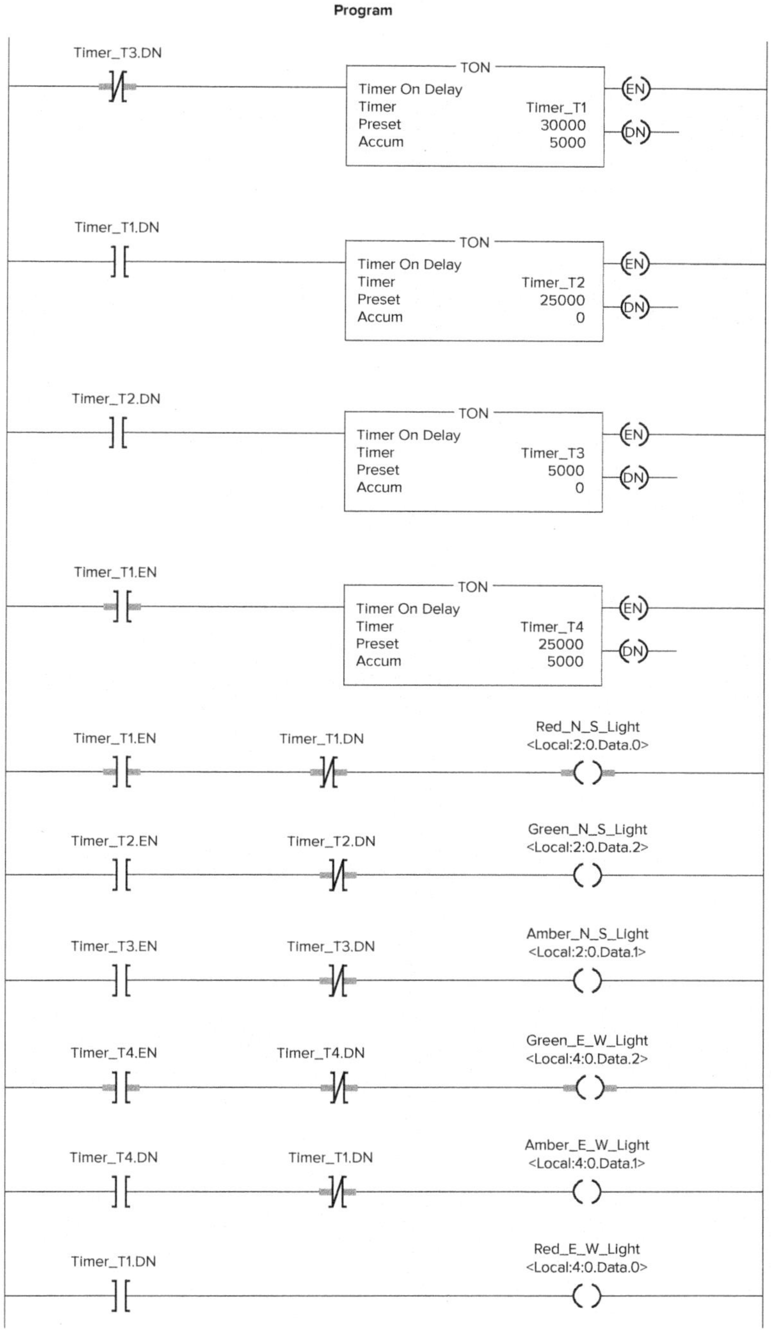

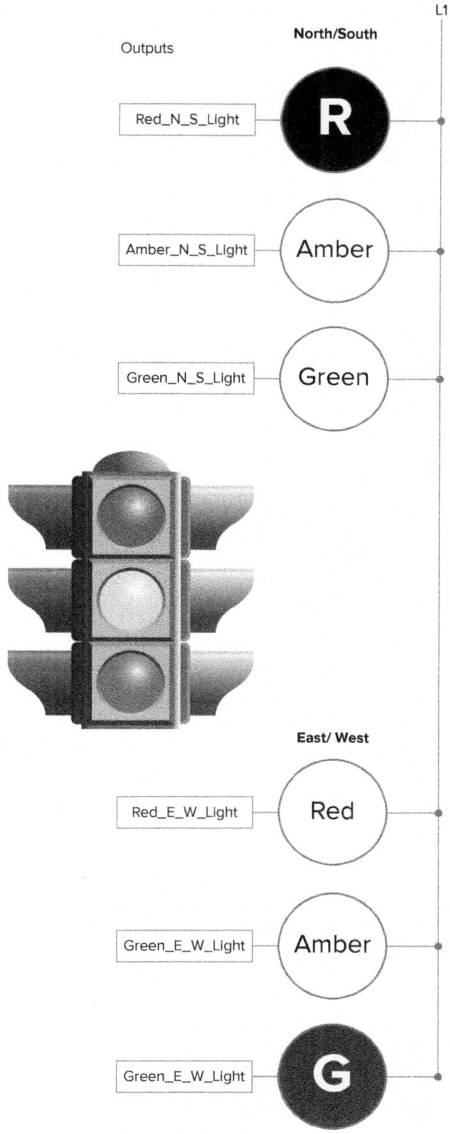

NAME ________________________________ DATE______________

7-21 Construct a program for the cascading TON timers program shown. Test your program for proper operation.

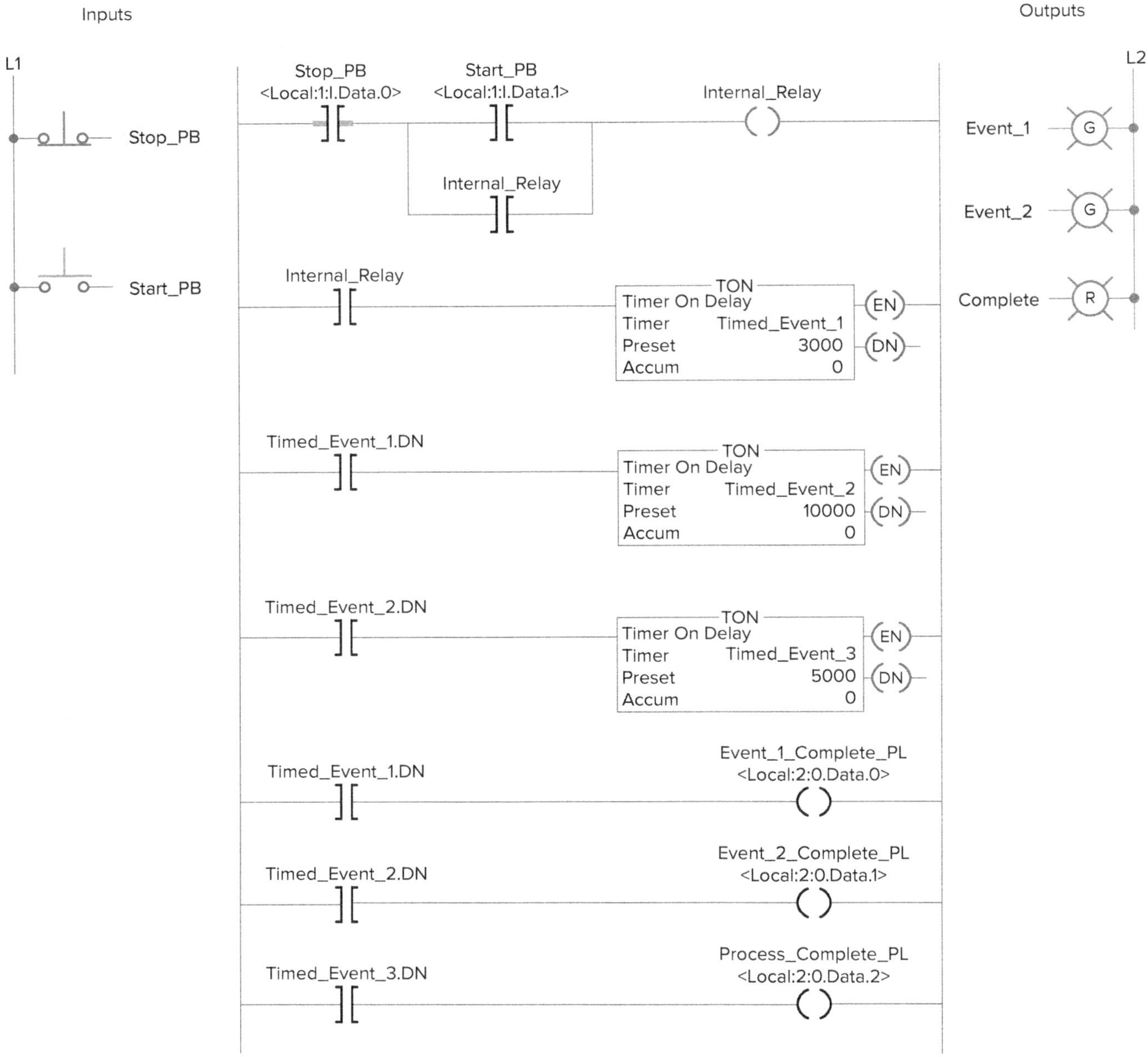

NAME ________________________________ DATE______________

7-22 Construct a program that will correctly execute the hardwired relay control circuit shown. Test your program for proper operation.

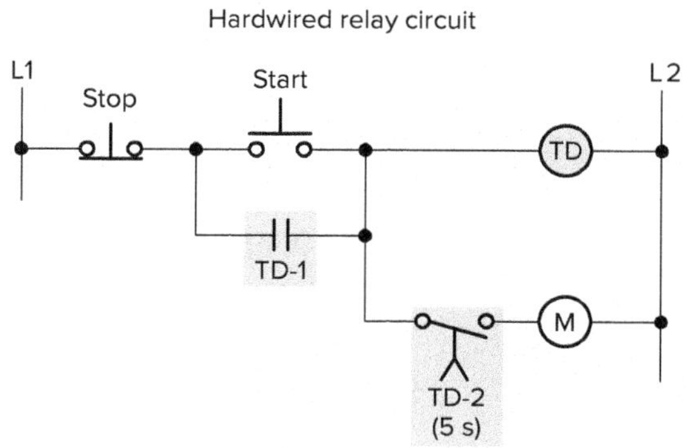

7-23 Construct a program for the retentive timer program shown. Operate your program and summarize its operation.

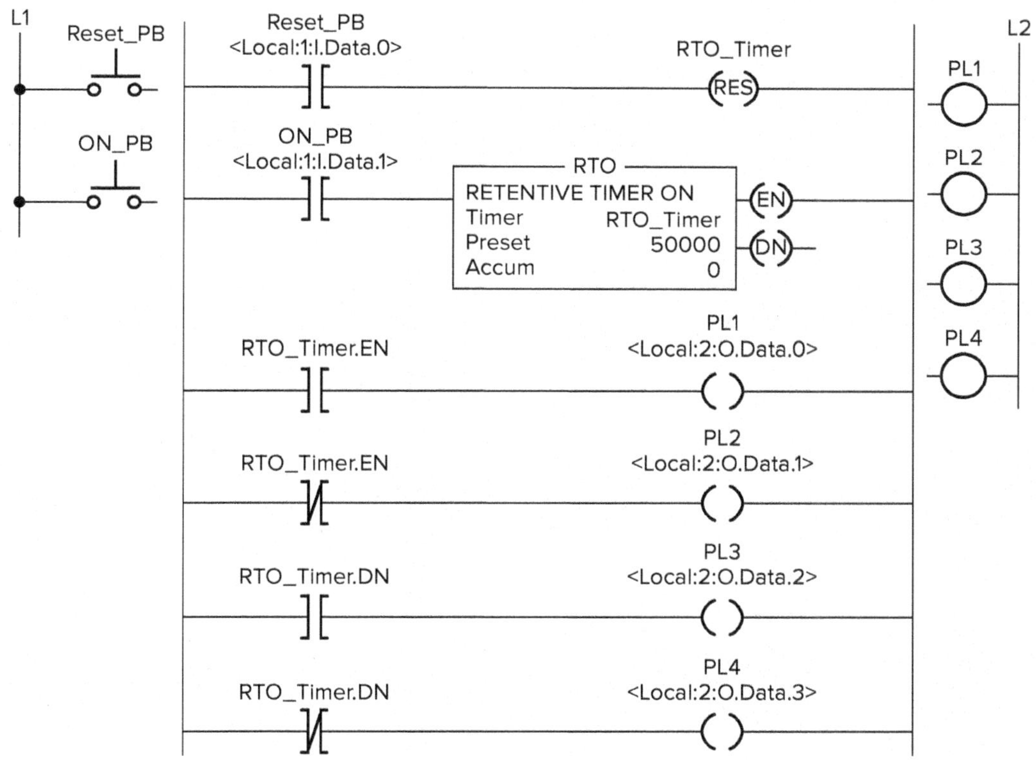

NAME ______________________________ DATE______________

7-24 Design a PLC program that will implement the hardwired circuit shown.

Download your program to the PLC and summarize its operation.

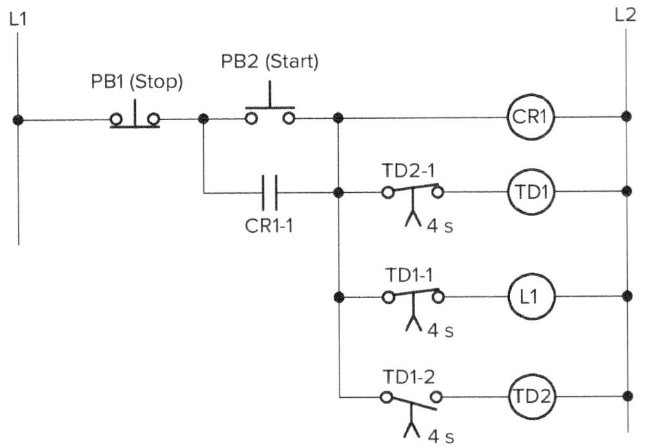

7-25 Design a PLC program that will implement the hardwired motor control circuit shown. Download your program to the PLC and confirm its operation. The desired operation of the process is summarized as follows:

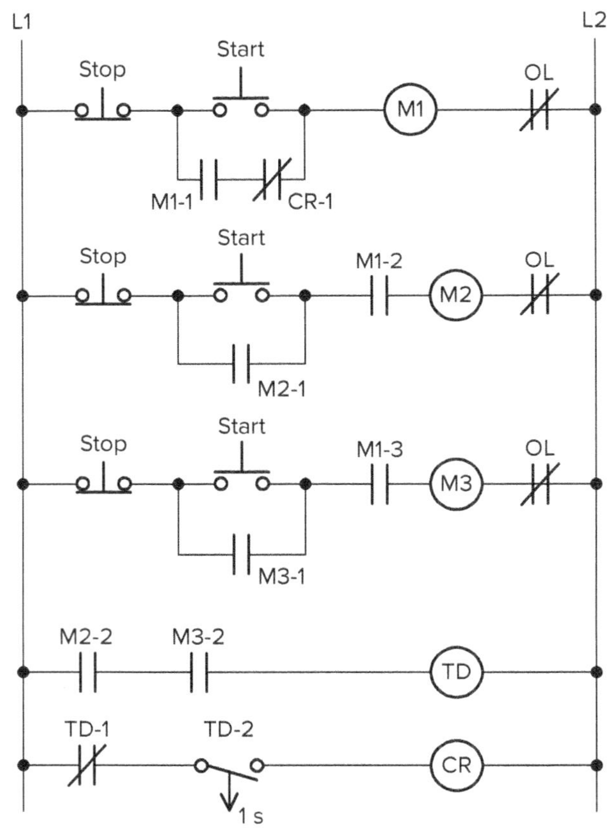

- The control process consists of three motors: M1, M2, and M3.
- The electrical control system is to be designed so that motor M1 must be running before motor M2 or M3 can be started.
- Each motor has its own start/stop pushbutton station.
- Both motors M2 and M3 can normally be stopped or started without affecting the operation of motor M1.
- However, if all three motors are running, the stopping of any one motor, for any reason, will automatically stop all three motors.

NAME ________________________________ DATE______________

7-26 Design a PLC program that will:

- Turn on pilot light PL1 10 seconds after switch S1 is turned on.
- Pilot light PL2 will come on 5 seconds after PL1 comes on.
- Pilot light PL3 will come on 8 seconds after PL2 comes on.
- Pressing pushbutton PB1 will reset all the timers but only if PL3 is on.

Download your program to the PLC and confirm its operation.

7-27 When a switch is turned on, PL1 goes on immediately and PL2 goes on 9 seconds later. Opening the switch turns both lights off. Design a program that implements this process. Download your program to the PLC and confirm its operation.

7-28 When a switch is turned on, PL1 and PL2 immediately go on. When the switch is turned off, PL1 immediately goes off. PL2 remains on for another 3 seconds and then goes off. Design a program that implements this process. Download your program to the PLC and confirm its operation.

7-29 When a switch is turned on, PL1 and PL2 immediately go on. PL1 turns off after 4 seconds. PL2 remains on until the switch is turned off. Turning the switch off at any time turns both lights off. Design a program that implements this process. Download your program to the PLC and confirm its operation.

7-30 Design a program that will implement the following control process. Download your program to the PLC and confirm its operation.

- A saw, fan, and lube pump all go on when a start button is pressed.
- Pressing a stop button immediately stops the saw but allows the fan to continue operating.
- The fan is to run for an additional 5 seconds after shutdown of the saw.
- If the saw has operated for more than 20 seconds, the fan should remain on until reset by a separate fan reset button.
- If the saw has operated less than 20 seconds, the lube pump should go off when the saw is turned off.
- However, if the saw has operated for more than 20 seconds, the lube pump should remain on for an additional 10 seconds after the saw is turned off.

NAME ______________________________ DATE______________

7-31 A lamp is required to turn on for 10 seconds each time a door is opened. Design a program that will implement this. Download your program to the PLC and confirm its operation.

7-32 A PLC program is required to control part of an industrial heating oven as follows:

- The system is activated by a Start button that seals in an internal control relay and is deactivated by pressing a Stop button.
- When the Start button is pressed, a TON timer will activate a warning buzzer which is on for 5 seconds and warns that the oven is about to start.
- After 5 seconds the buzzer stops and the heating coils for the oven are turned on for 10 minutes (600 seconds).

Design a program that will implement this. Download your program to the PLC and confirm its operation.

NAME ______________________ DATE ______________

CHAPTER 8

Programming Counters

8-1 Construct the up-counter PLC program shown. Test your program for proper operation.

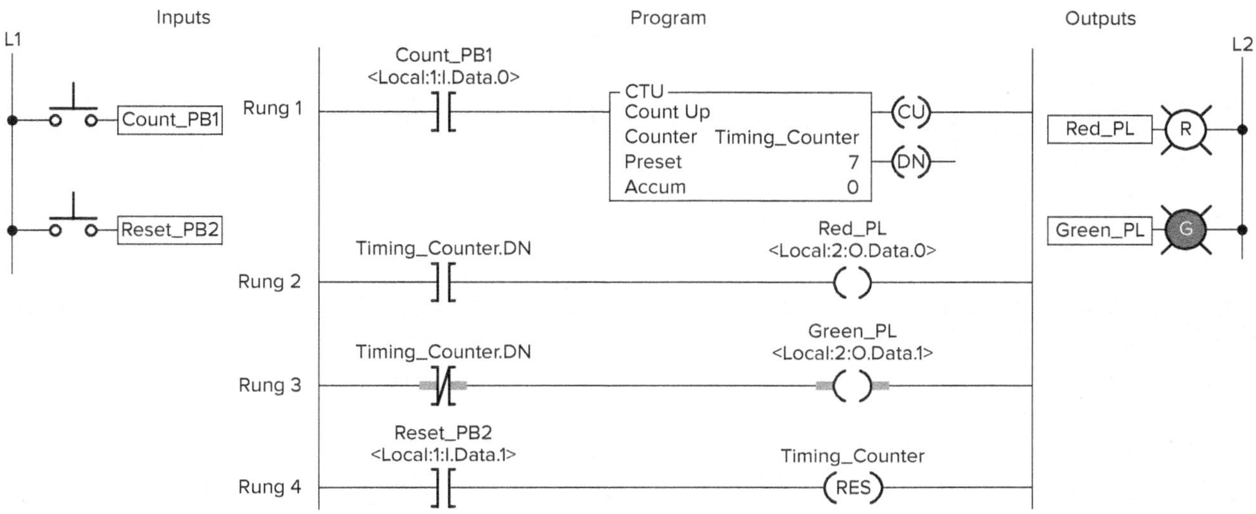

8-2(a) Construct the program to monitor the number of motor start/stop operations shown. Test your program for proper operation.

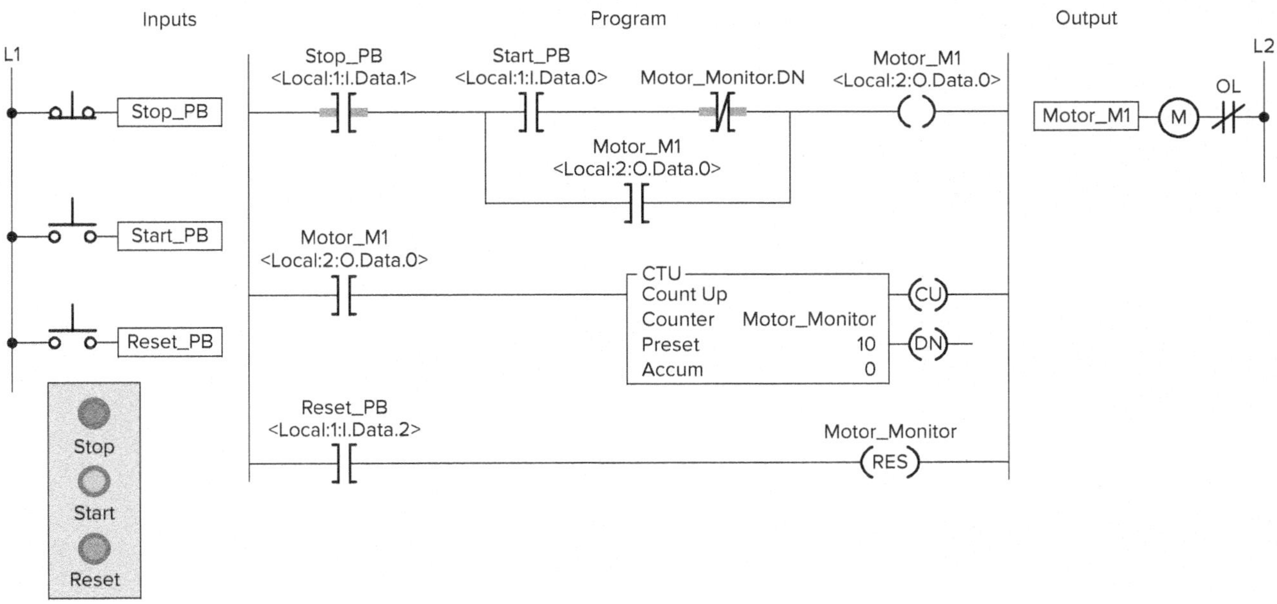

NAME ________________________________ DATE______________

8-2(b) Add a green pilot light to indicate when the count is less than 10 and a red pilot light to indicate when the count has reached 10.

8-3 Construct the PLC cans package process shown. Test your program for proper operation.

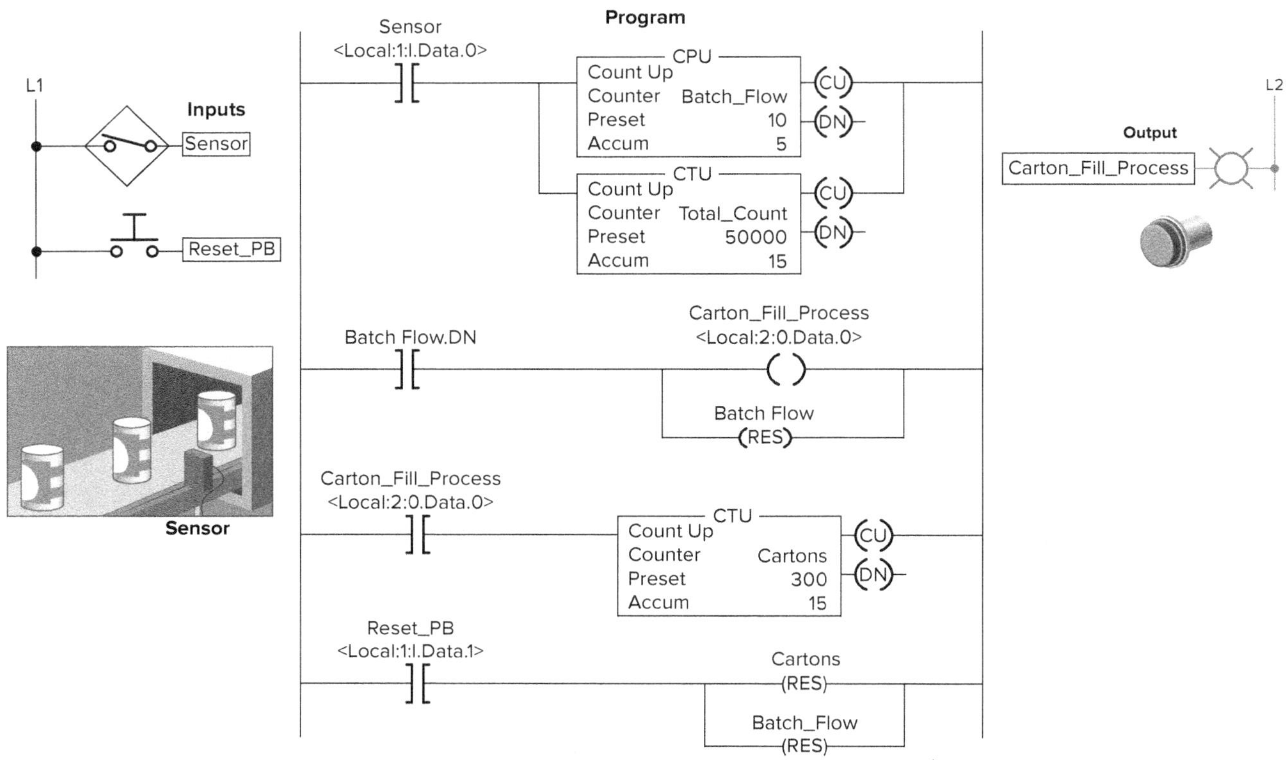

8-4 Construct the program for removal of containers from a conveyor line shown. Test your program for proper operation as outlined in the text.

NAME ____________________ DATE ____________

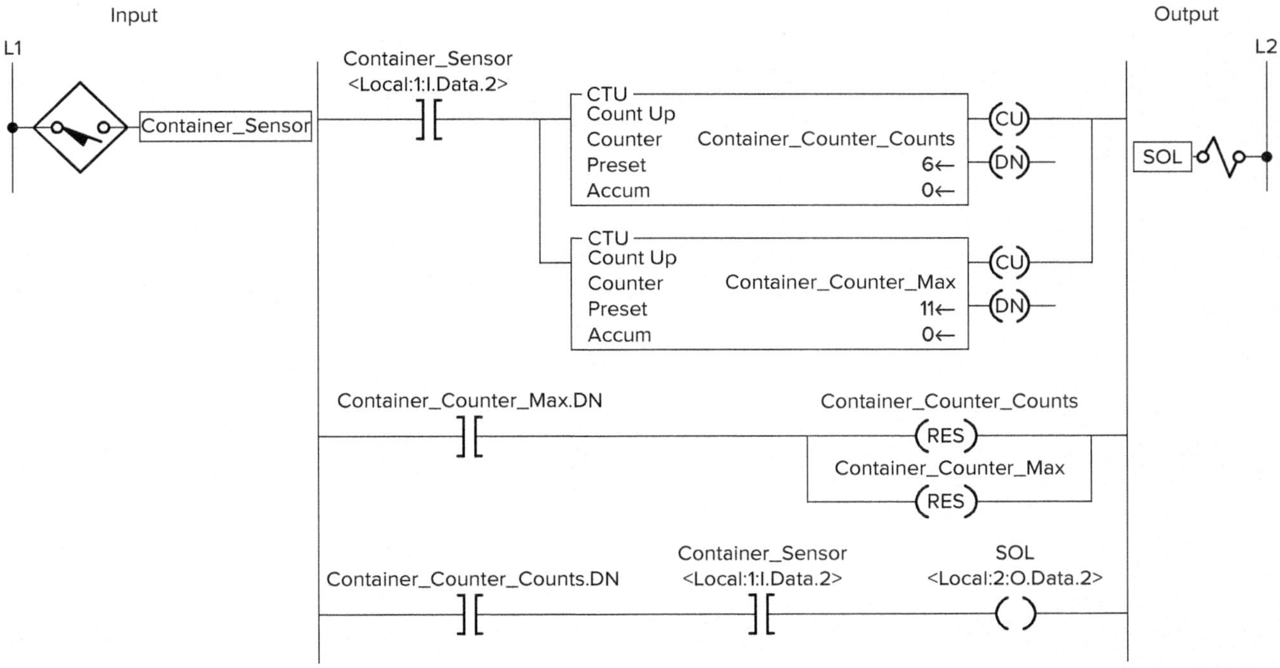

8-5 Construct the PLC program that uses a one-shot instruction to trigger count pulses. Test your program for proper operation.

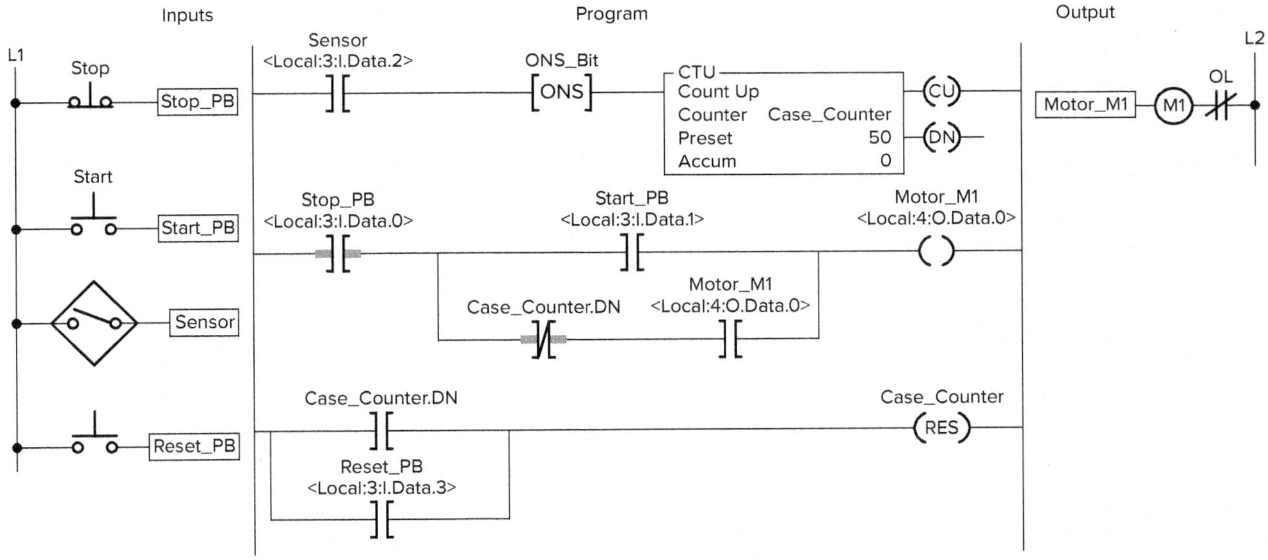

NAME ______________________ DATE______________

8-6 Construct the PLC one-shot program used to send data to an LED display. Test your program for proper operation.

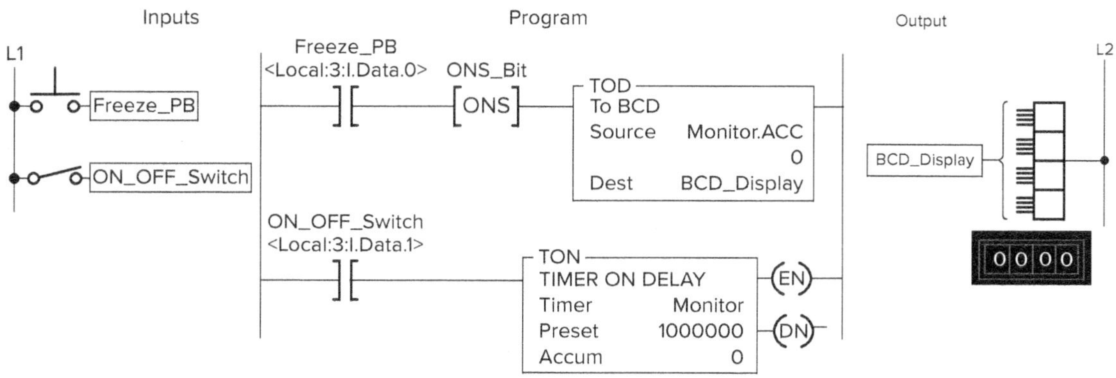

8-7 Construct the PLC program that uses a one-shot instruction to trigger count pulses. Test your program for proper operation.

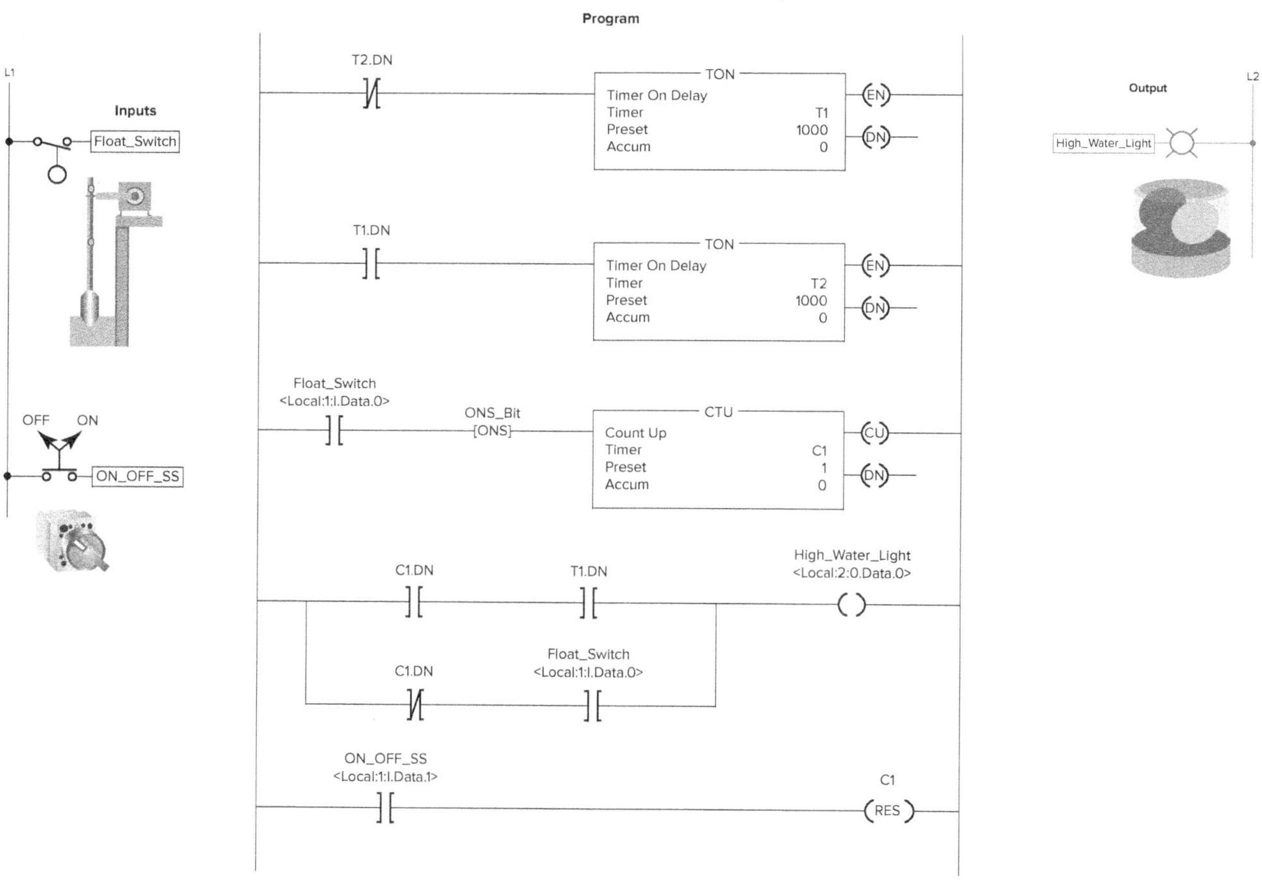

NAME ____________________ DATE__________

8-8 Construct the combined CTU and CTD program. Test your program for proper operation.

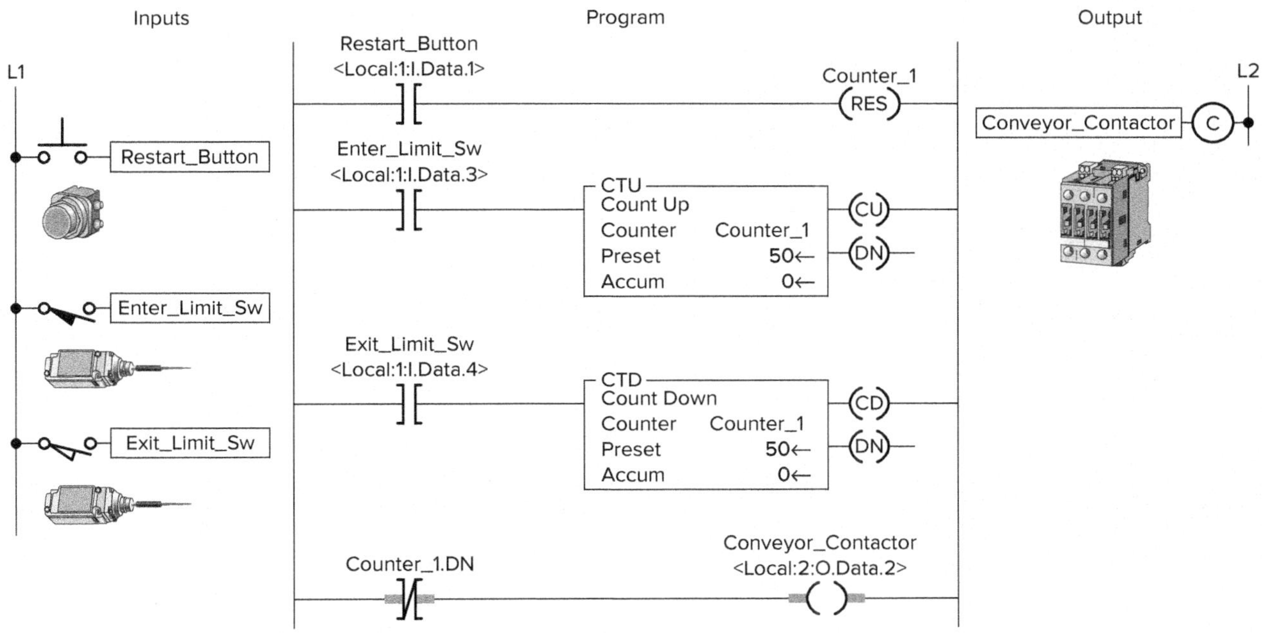

8-9(a) Construct the parking lot counter program shown. Test your program for proper operation.

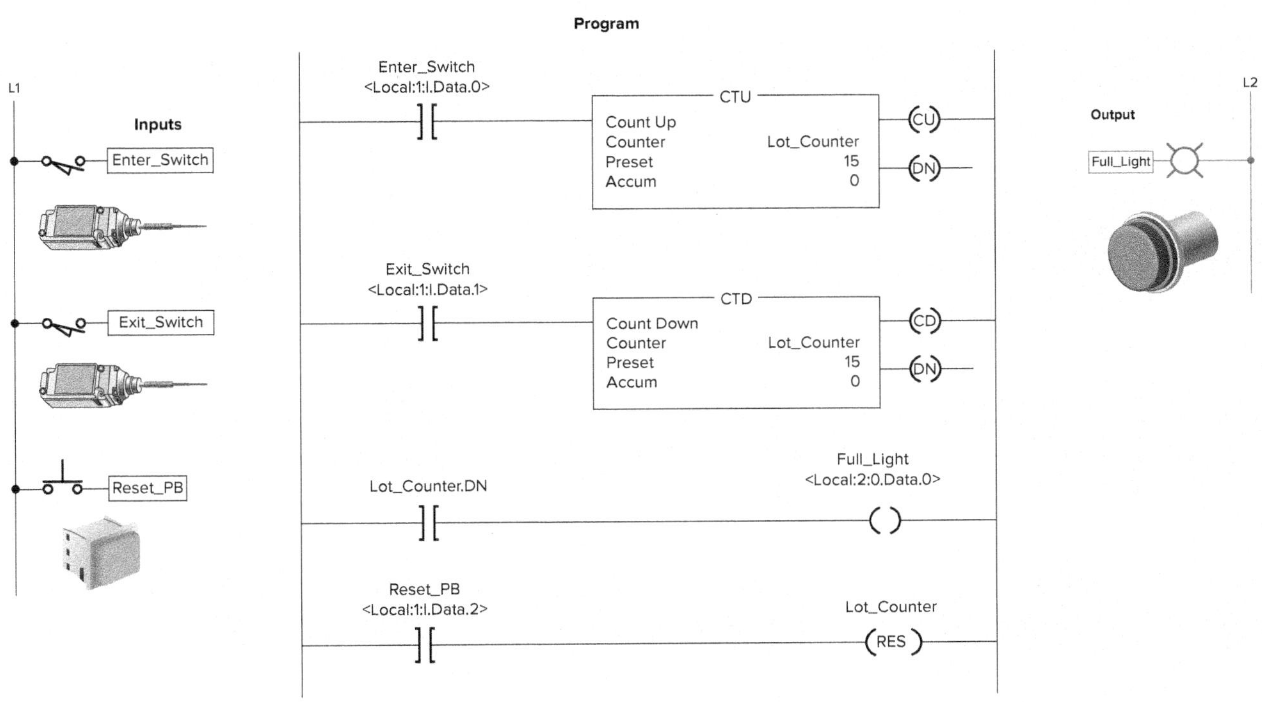

NAME ______________________________ DATE______________

8-9(b) Modify your program to include a warning pilot light that comes on when the lot is 80% full.

8-10 Construct the process monitoring up/down counter program shown. Test your program for proper operation.

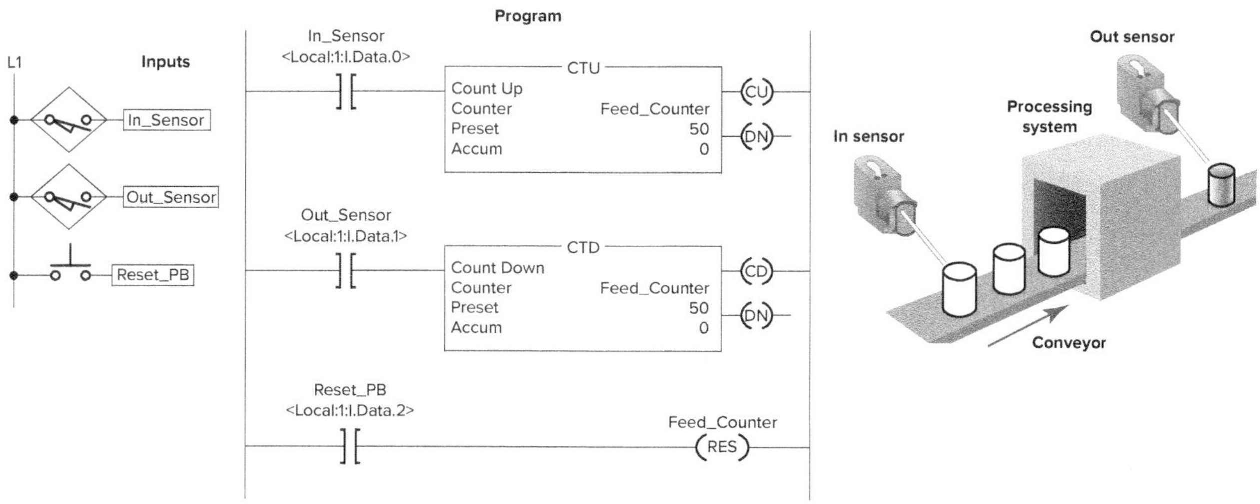

8-11 Construct the program for cascading of counters shown. Test your program for proper operation.

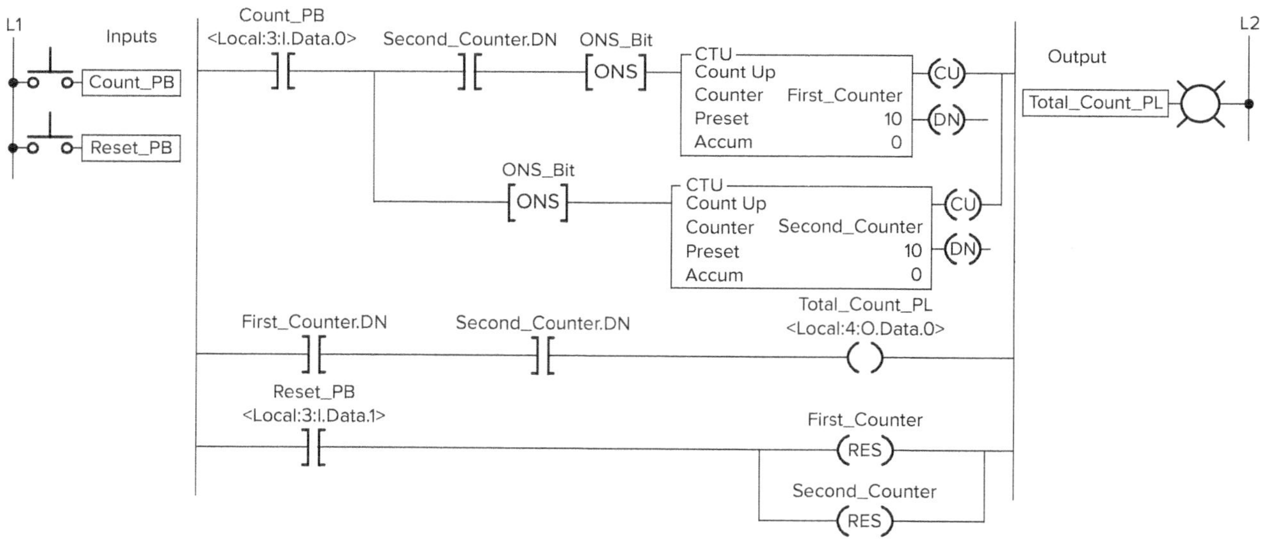

NAME ______________________________ DATE______________

8-12 Construct the 24-hour clock program shown. Test your program for proper operation.

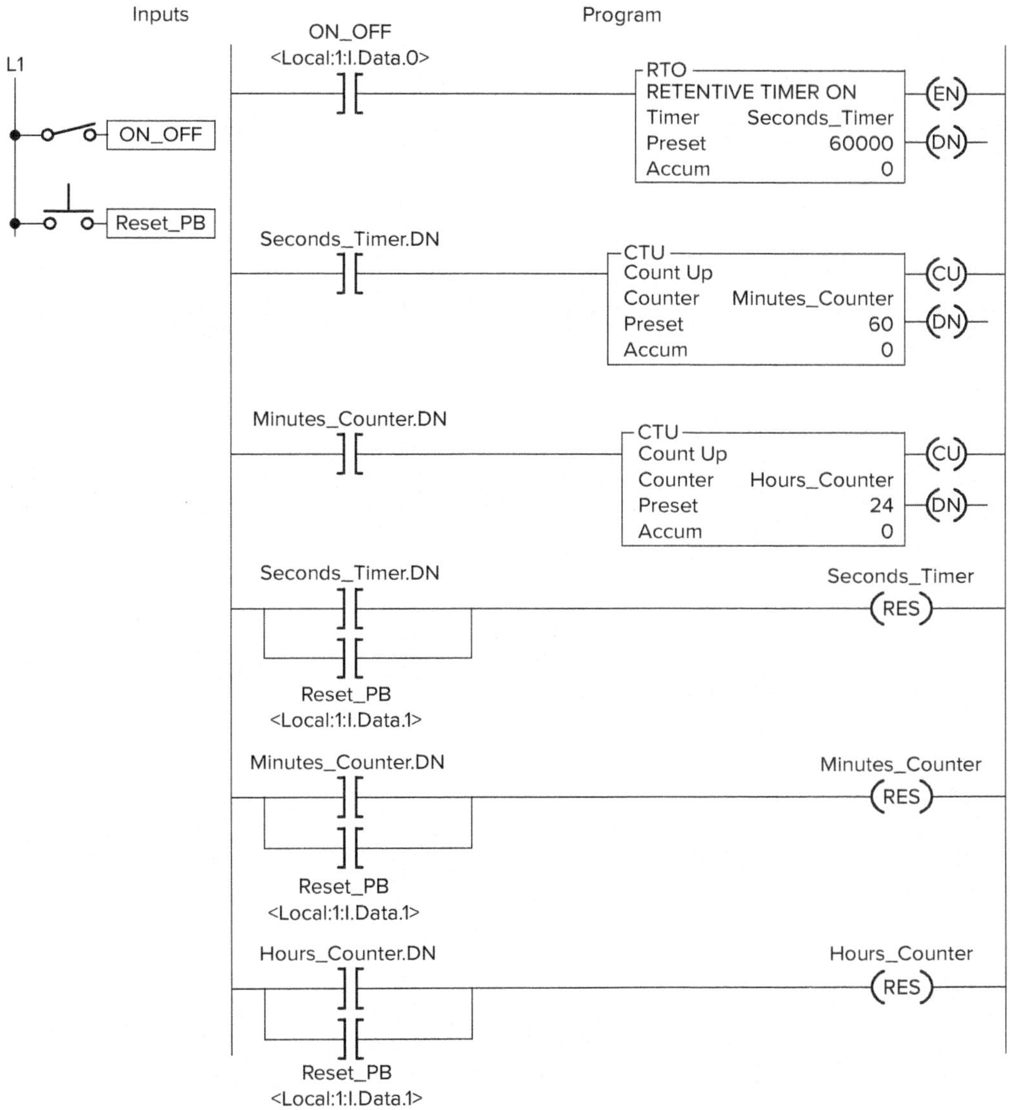

NAME ________________________________ DATE________________

8-13 Construct the program shown for monitoring the time of an event. Test your program for proper operation.

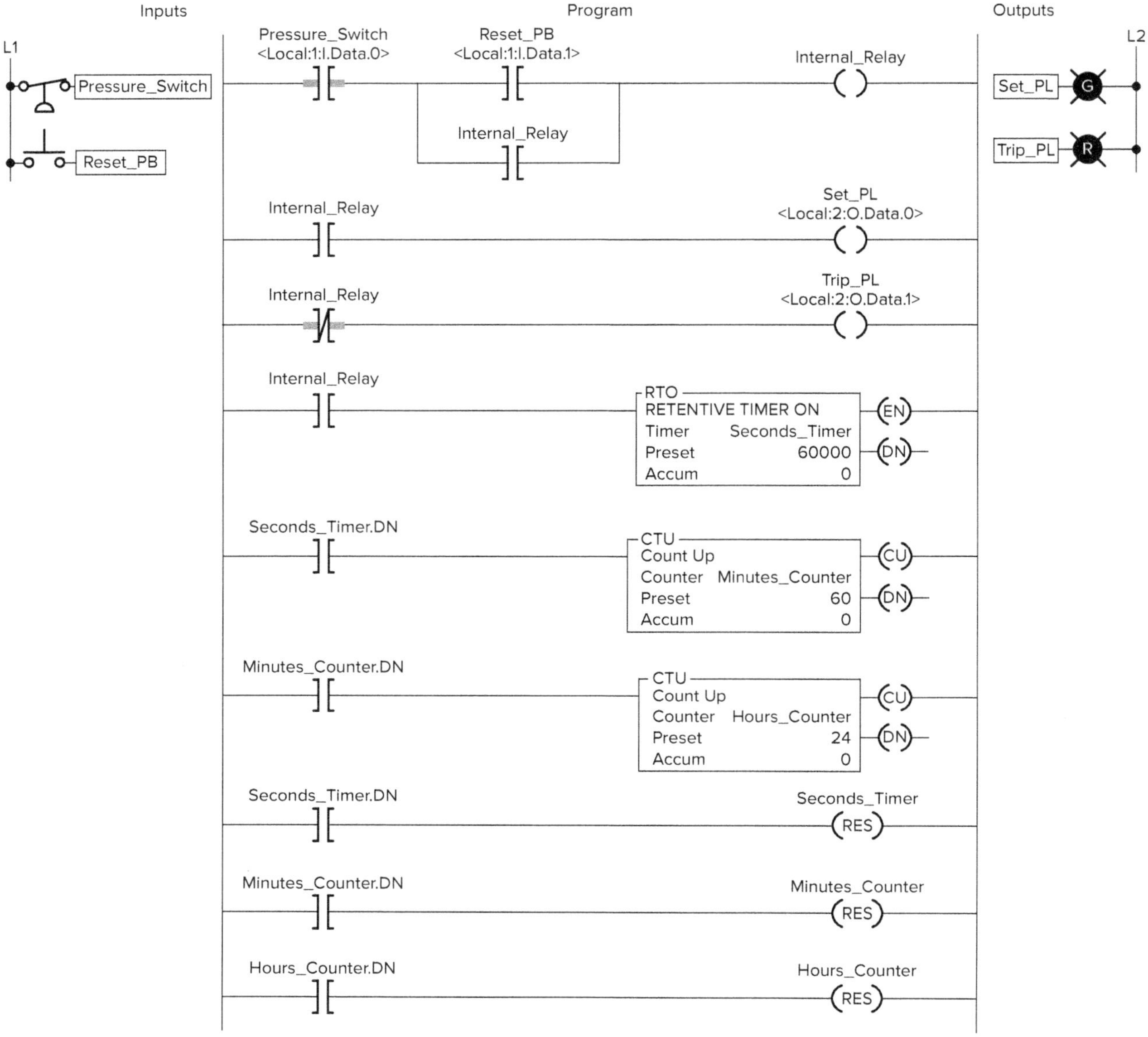

NAME ______________________________ DATE ______________

8-14 Construct the cut-to-length measurement program shown. Test your program for proper operation.

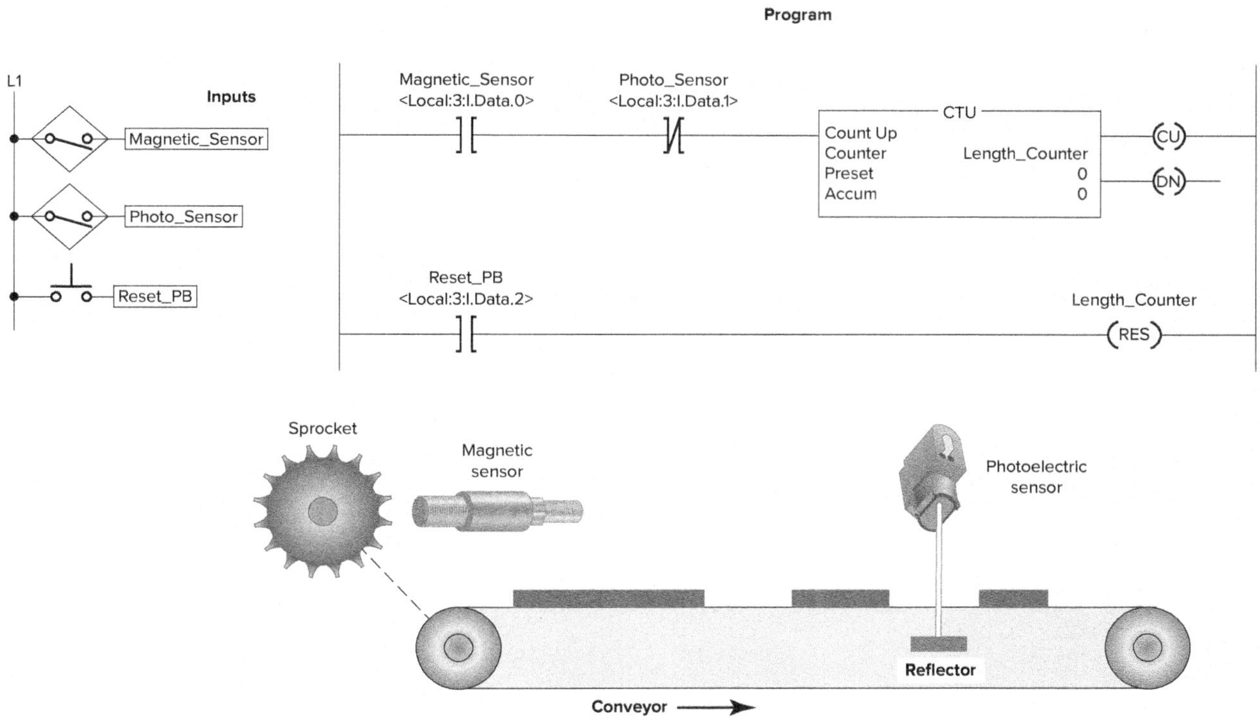

8-15 Construct the automatic stacking program shown. Test your program for proper operation.

NAME ______________________________ DATE ______________

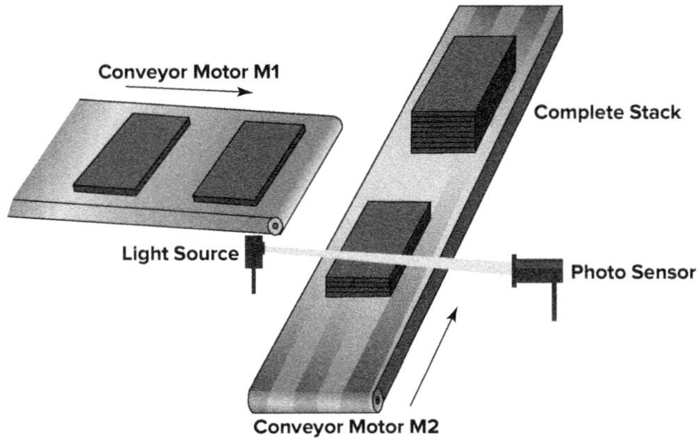

Program

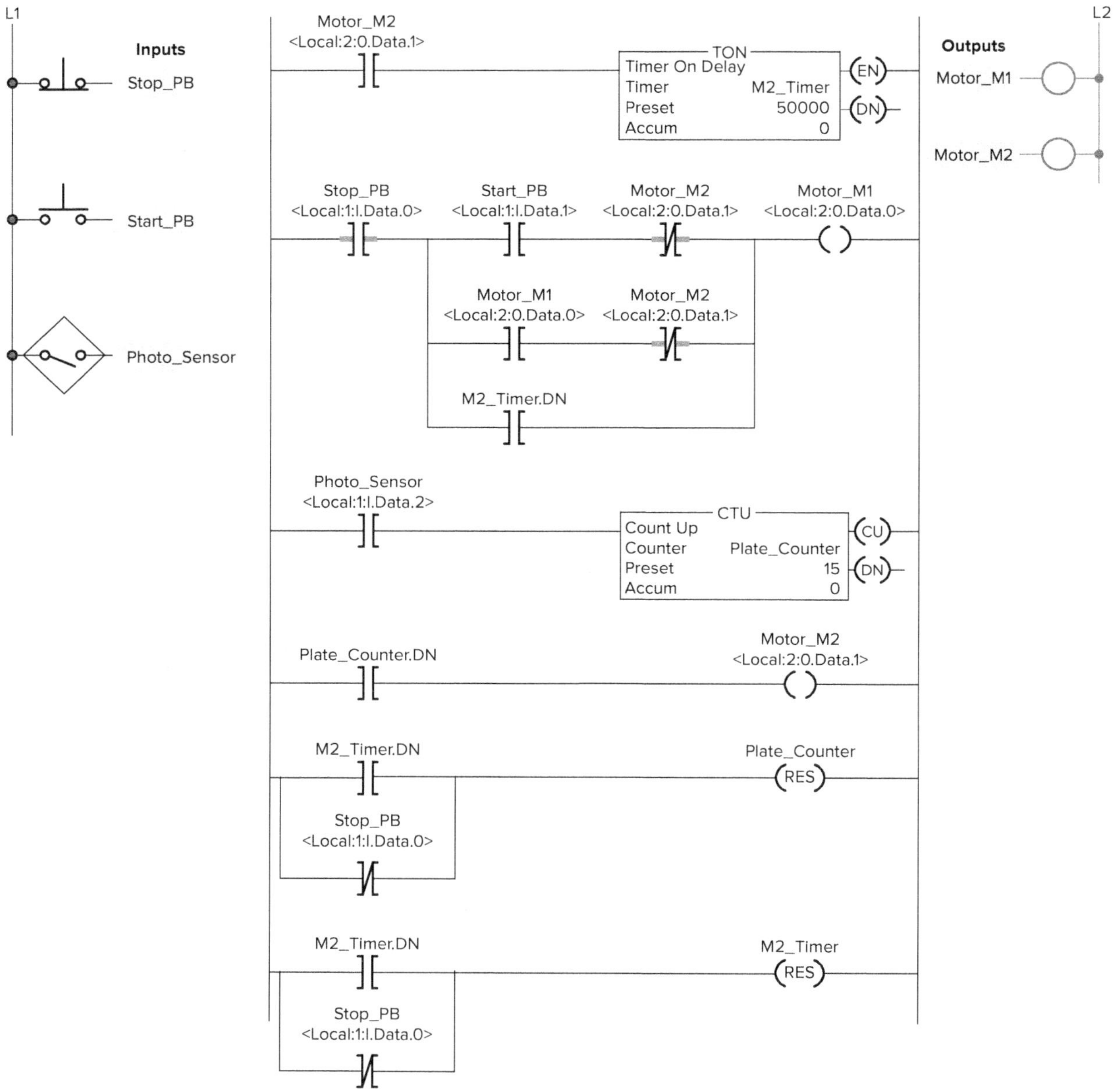

NAME ________________________________ DATE ______________

8-16 Construct the motor lock-out program shown. Test your program for proper operation.

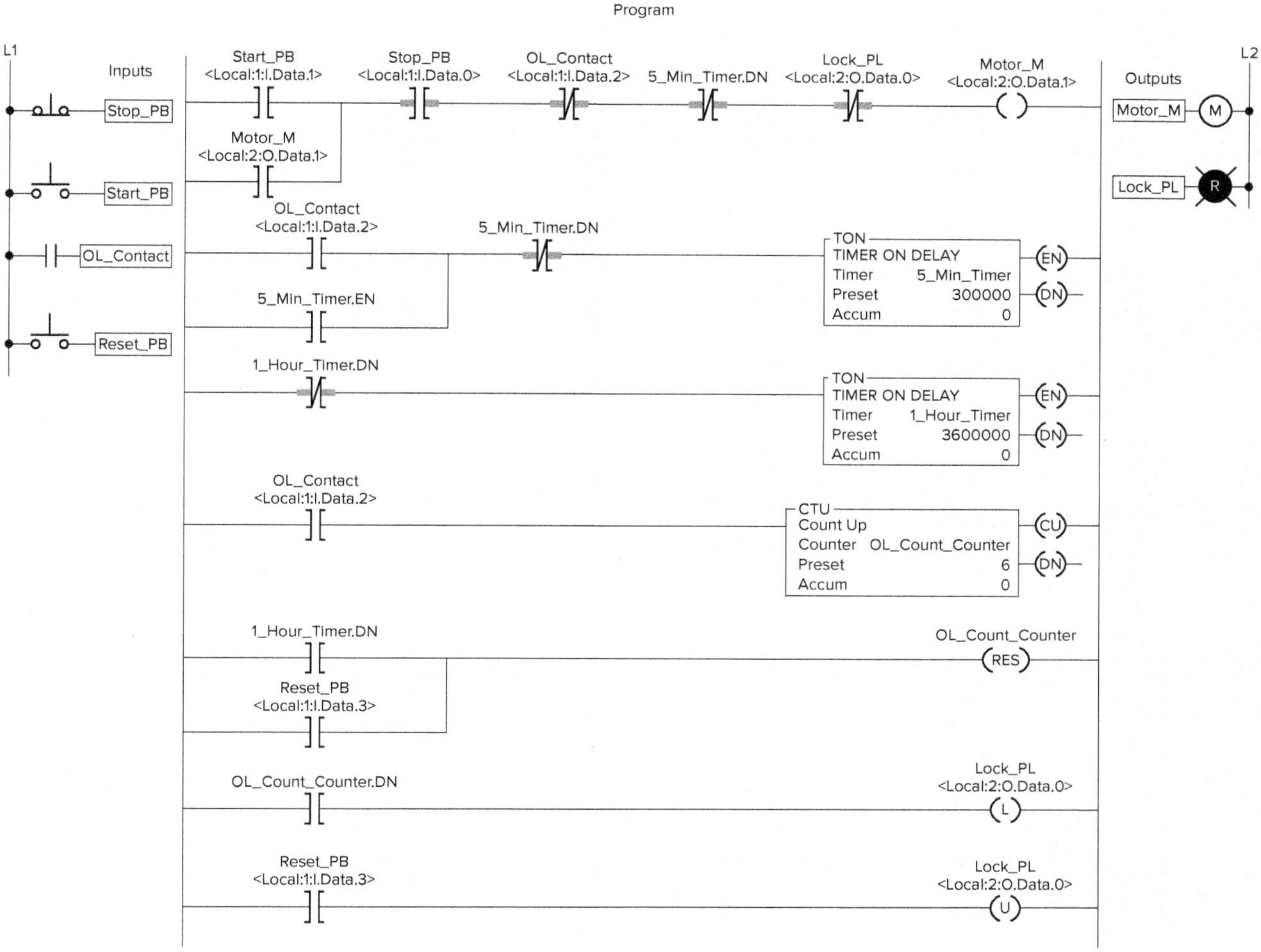

NAME ______________________________ DATE ______________

8-17 Construct the bottle flow rate program shown. Test your program for proper operation.

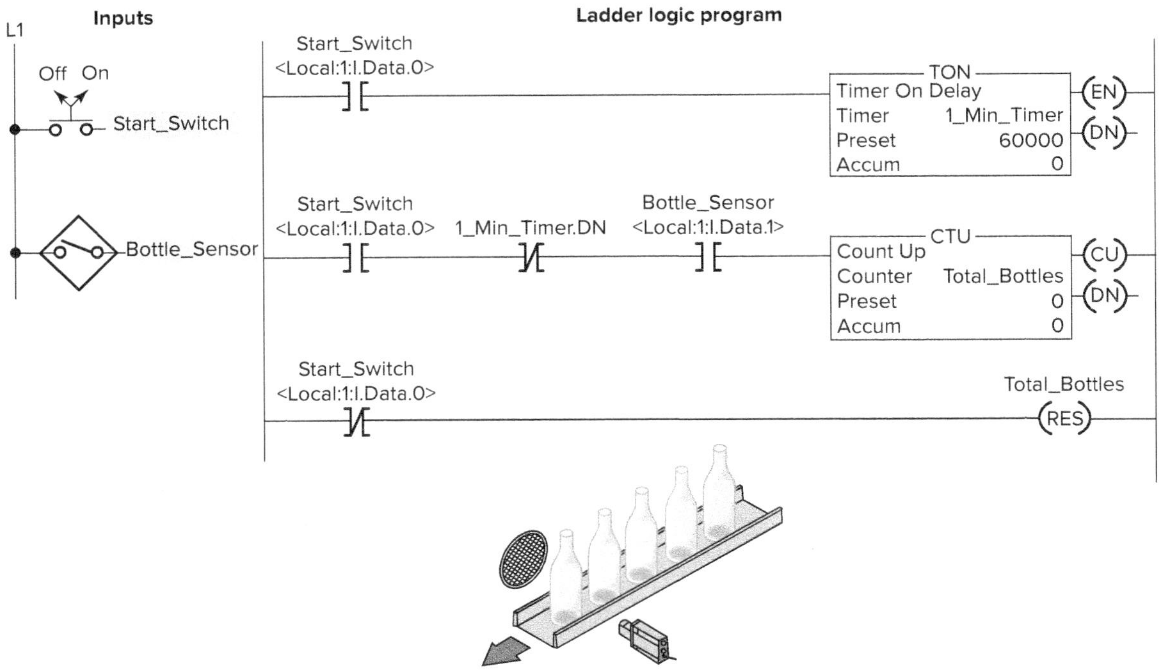

8-18 Construct the extremely long time-delay program shown. Test your program for proper operation.

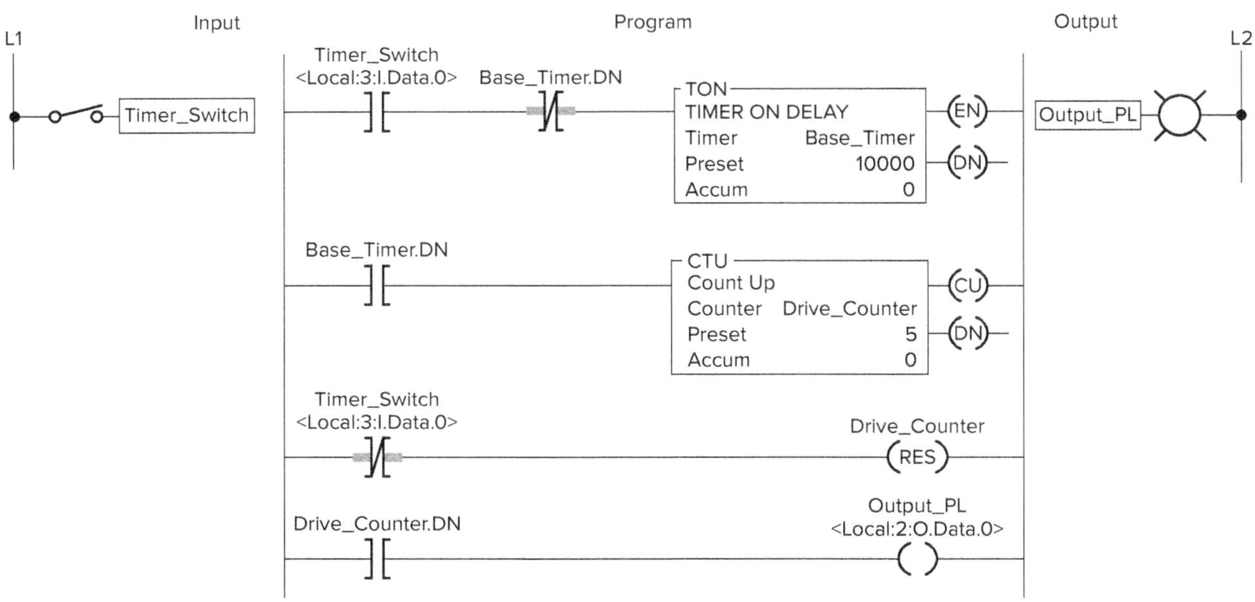

NAME ____________________ DATE____________

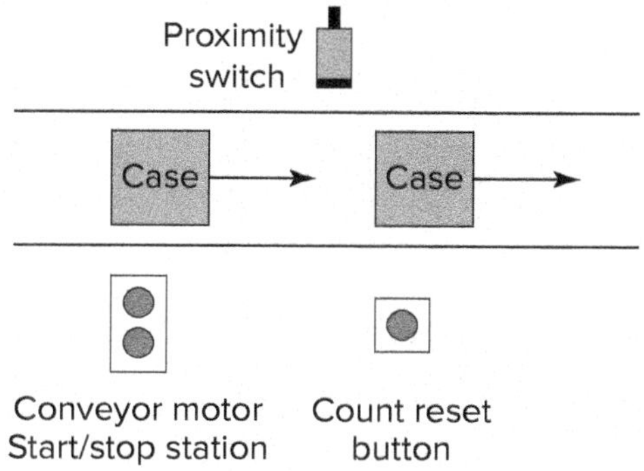

8-19 Design a PLC program that will correctly execute the conveyor motor control process illustrated. Download the program to the PLC and verify its operation. The required operational sequence is summarized as follows:

- The start button is pressed to start the conveyor motor.
- Cases move past the proximity switch and increment the counter's accumulated value.
- After a count of 50, the conveyor motor stops automatically and the counter's accumulated value is reset to zero.
- The conveyor motor can be stopped and started manually at any time without loss of the accumulated count.
- The accumulated count of the counter can be reset manually at any time by means of the counter reset button.

8-20 Design a PLC program that will latch on an output, PL1, after an input, PB1, has cycled on 20 times. When the count of 20 is reached, the counter will reset itself automatically. PB2 will unlatch PL1. Download the program to the PLC and verify its operation.

8-21 Design a PLC program that will correctly execute the packaging process shown. Download the program to the PLC and verify its operation. The operational sequence is summarized as follows:

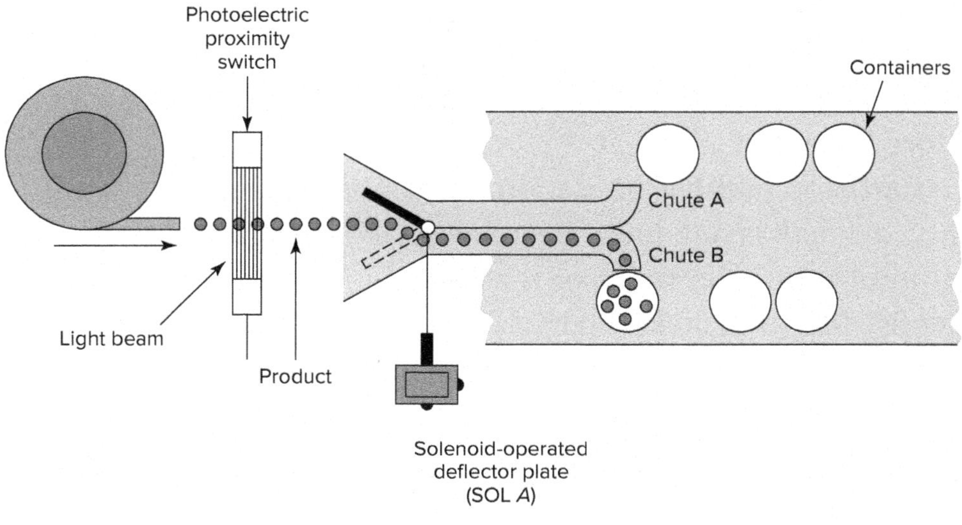

- The purpose of this process is to deposit 50 pieces of the product in each container.
- The process is set in operation by pressing a start pushbutton.
- As the product passes through the light beam, it is detected by the photoelectric proximity switch and counted by the PLC counter.
- When the count reaches 50, the solenoid-operated deflector plate (SOL A) energizes to channel the product from chute A to chute B.
- The counter is reset automatically for the next count of 50.
- When the second count of 50 is reached, the solenoid-operated deflector plate de-energizes to channel the product back into chute A, and so on.
- Provisions must be made for stopping the process at any time and manually resetting the accumulated value of the counter to any number.

8-22 Design a PLC program to operate a pilot light according to following sequence. Download the program to the PLC and verify its operation.

- A momentary pushbutton is pressed to start the sequence.
- The light is switched on and remains on for 5 seconds.
- The light is then switched off and remains off for 5 seconds.
- A counter is incremented by 1 after this sequence.
- The sequence then repeats for a total of five counts.
- After the fifth count, the sequence will stop and the counter will be reset to zero.

8-23 Create a PLC program to alternate the use of two tank fill pumps so that they both get the same amount of usage over their lifetime. Download the program to the PLC and verify its operation. The control process is summarized as follows:

- A start/stop pushbutton station is provided for control of the two fill pump motors P1 and P2.
- The start/stop pushbutton station is operated to control pump P1.
- When the tank is full, drain pump motor P3 is started automatically and runs until the low-level sensor is actuated.
- After five fillings of tank by pump P1, control automatically shifts to pump P2.
- Operation of the start/stop pushbutton now controls pump P2.
- After five fillings of the tank by pump P2, the sequence is repeated.

NAME ______________________________ DATE______________

8-24 The PLC program shown is an example of a bottling operation where 5 out of every 10 bottles are removed as they pass by on a conveyor. Design and test a program which will emulate the operation of the program summarized as follows:

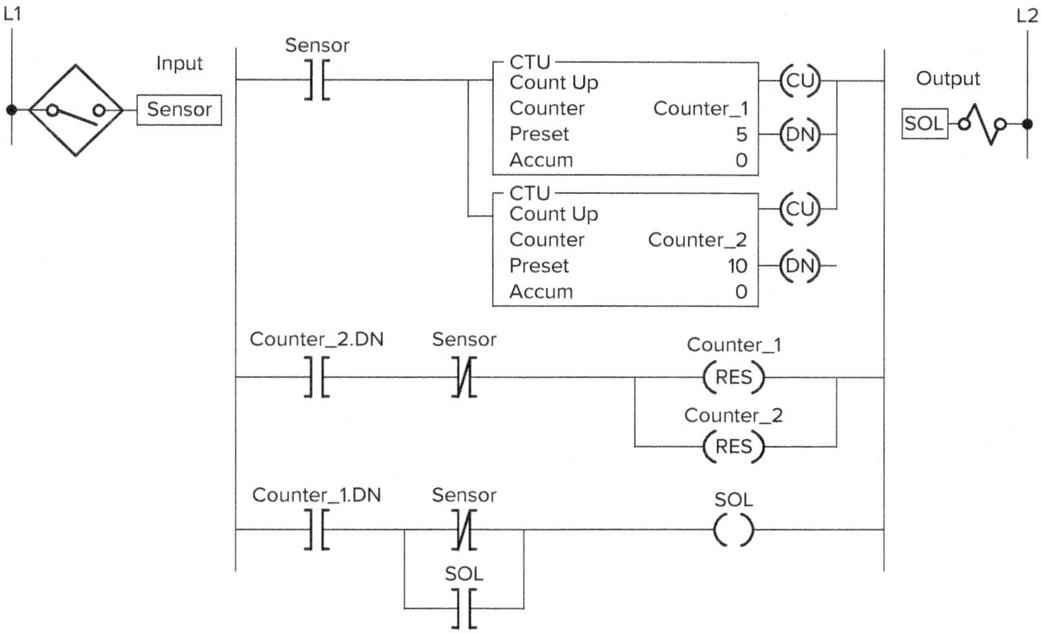

- Counter_1 Preset value is set for 5.
- Counter_2, Preset value is set for 10.
- As each bottle is detected by the main conveyor Sensor, both counters increment their value by 1.
- When the 6th bottle is detected, the diverter gate solenoid (SOL) is activated and diverts bottles 6 through 10 from the main conveyor.
- The diverter gate is then switched off to allow bottles 11 through 15 to pass by on the main conveyor.

NAME ________________________________ DATE______________

CHAPTER 9

Program Control Instructions

9-1 Construct the PLC Master Control Reset (MCR) program shown. Verify that when the MCR instruction is de-energized, all nonretentive outputs de-energize and all retentive outputs remain in their last state.

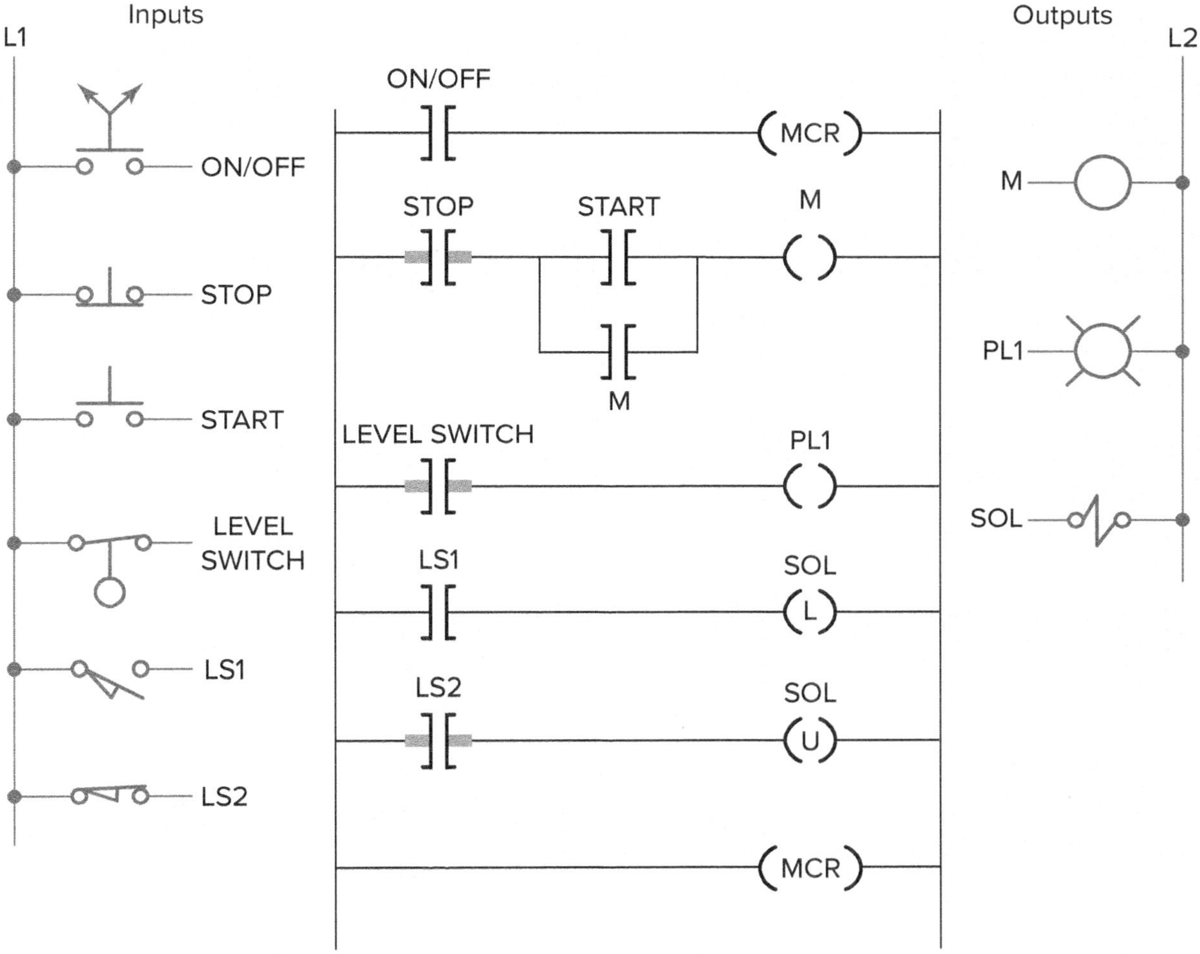

NAME ____________________ DATE ____________

9-2 Construct the PLC Master Control Reset (MCR) fenced zone program shown. Test your program for proper operation.

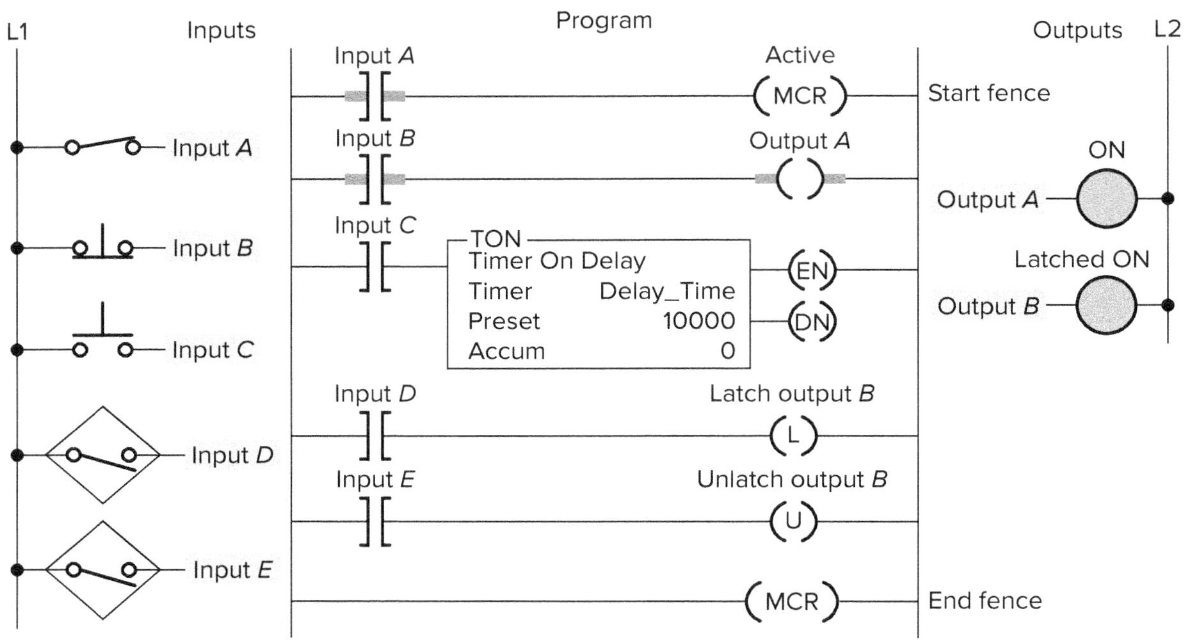

9-3 Construct the PLC Jump (JMP) program shown. Test your program for proper operation.

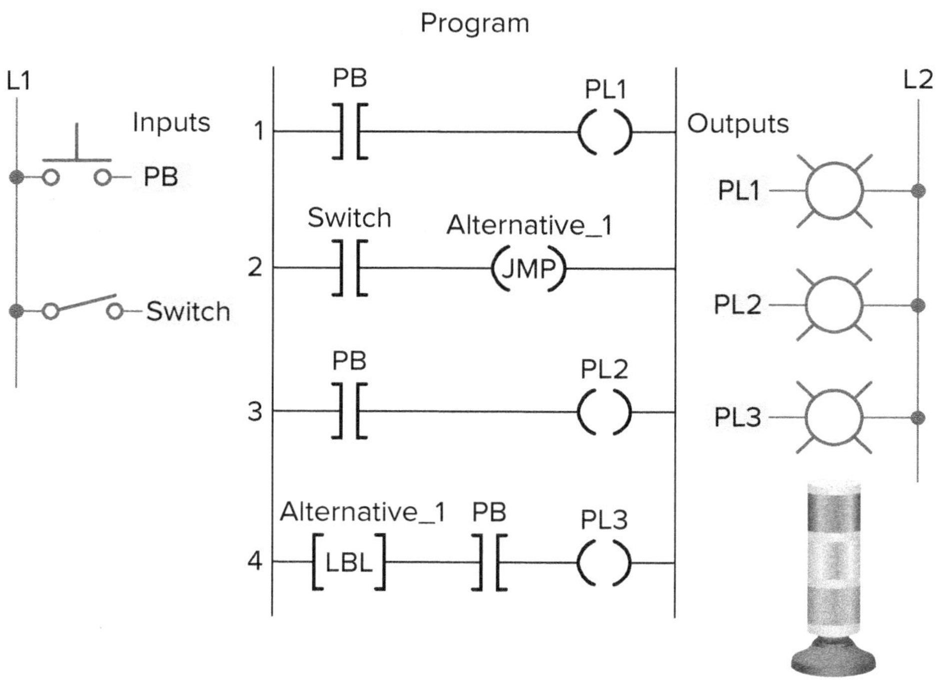

NAME ____________________________________ DATE________________

9-4 Construct the PLC jump-to-label program shown.
Test your program for proper operation.

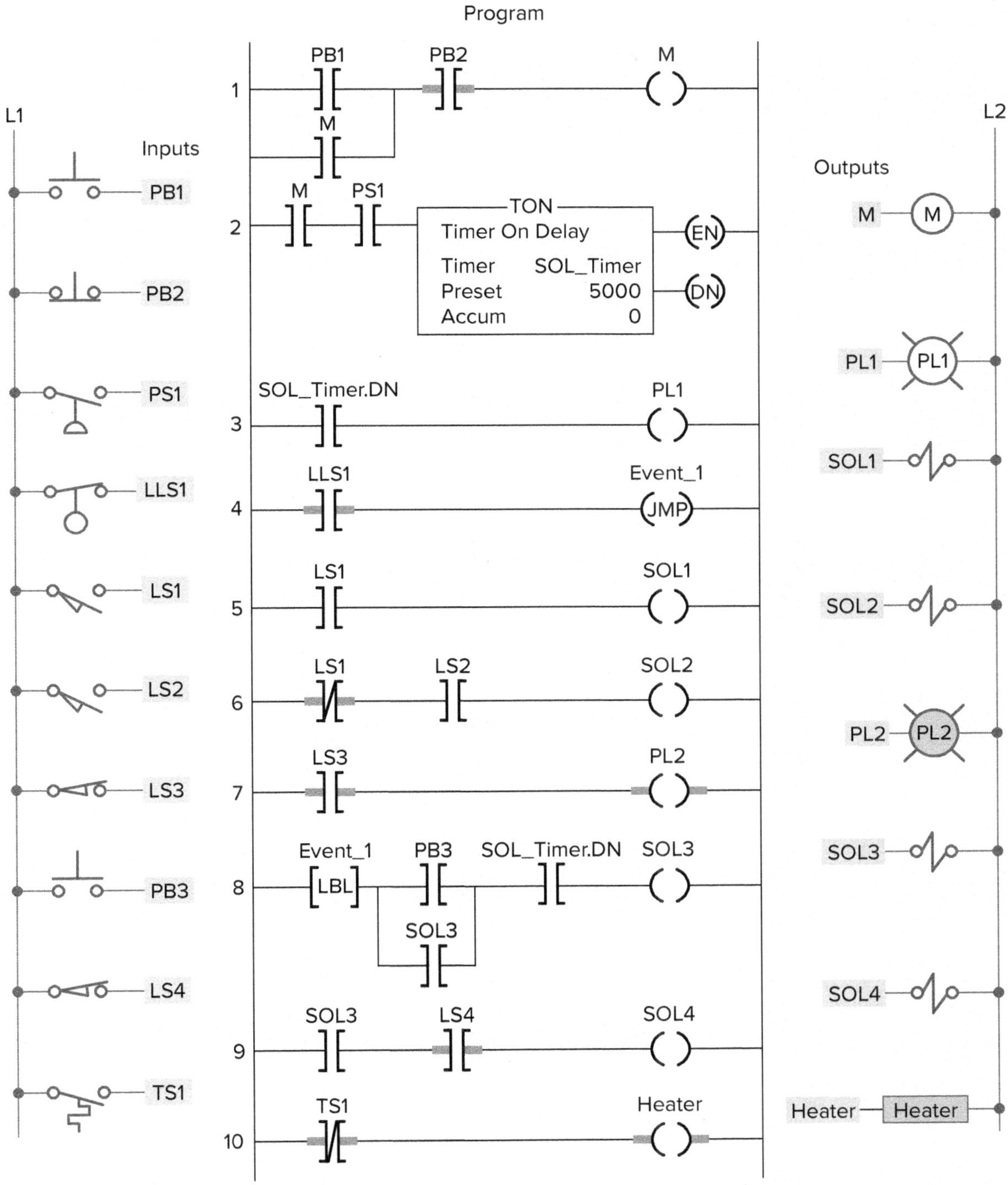

NAME ________________________________ DATE______________

9-5 Construct the PLC Flashing Pilot Light Subroutine program shown. Test your program for proper operation.

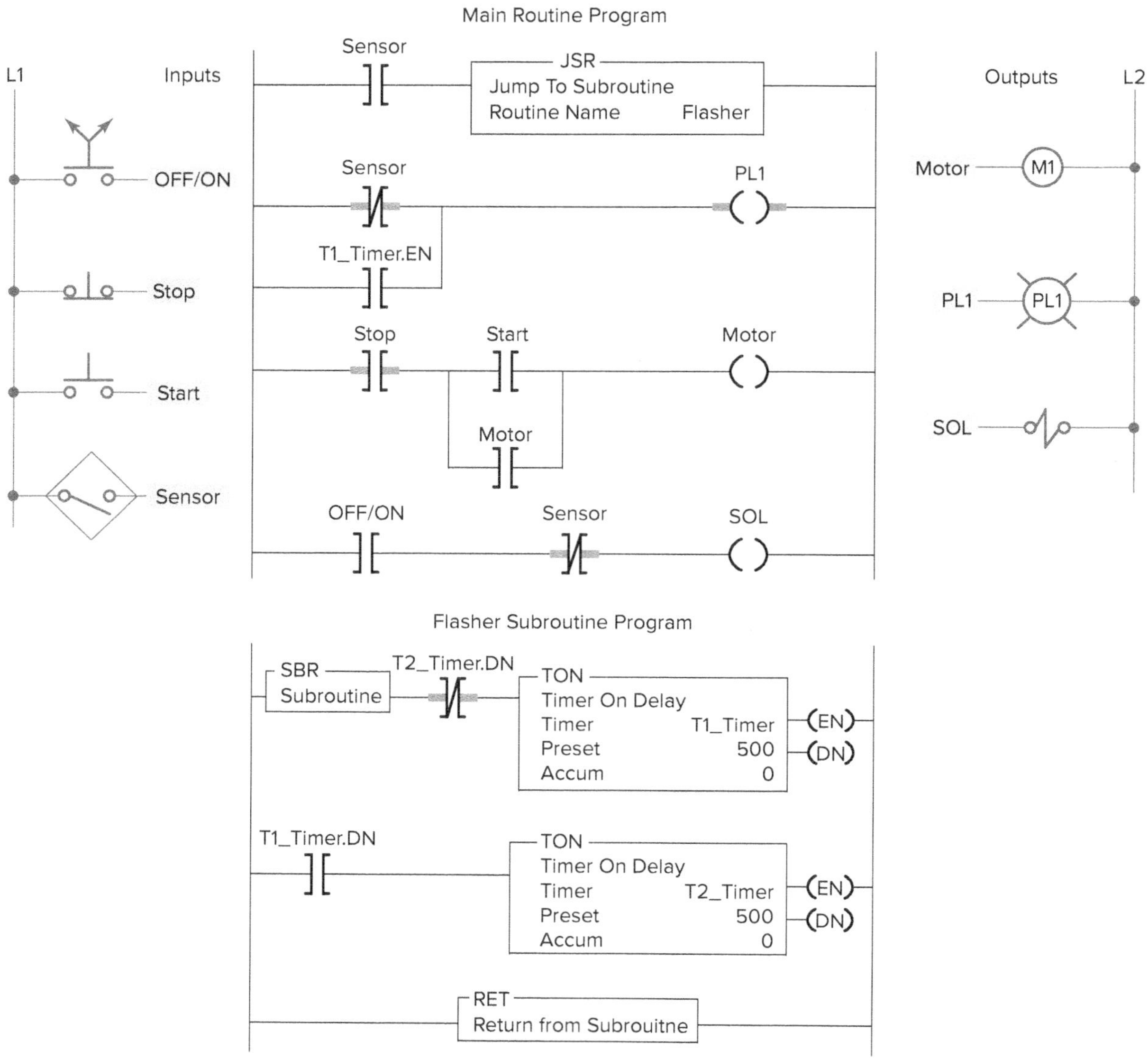

NAME ________________________________ DATE______________

9-6 Construct and download the program shown to the PLC. With reference to the text, operate the switch and make note of the change in normal operation for each of the following forcing procedures:

- Forcing the switch On
- Forcing the switch Off
- Forcing PL1 On
- Forcing PL1 Off

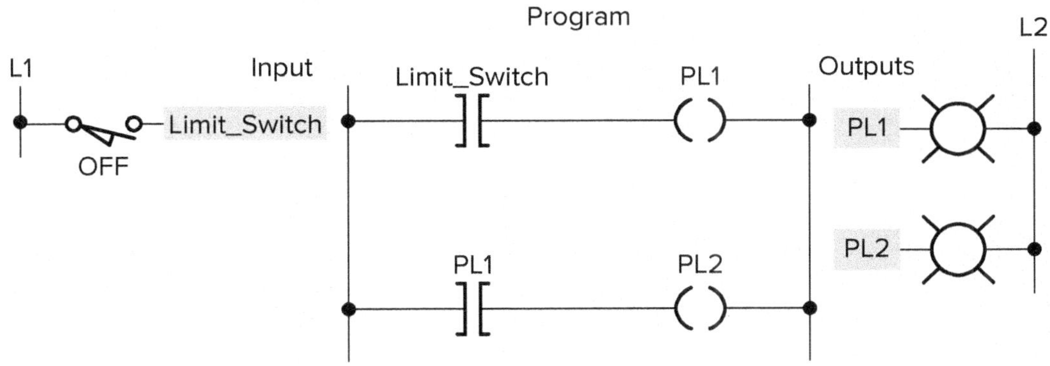

9-7 Construct and download the program shown to the PLC. Simulate:

- How an open in the Stop pushbutton will fail to de-energize the output.
- Replace the normally open Stop pushbutton with a normally closed type, and modify the program so that the circuit operates properly. Simulate how an open in the Stop pushbutton will automatically de-energize the output.

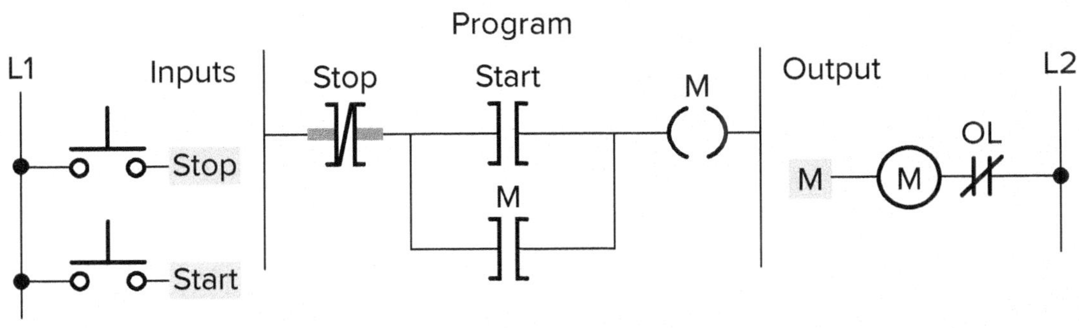

NAME ____________________ DATE ____________

9-8 Construct and download the PLC Pilot Lights Subroutine program shown. Compare the operation of the program when switch S2 is open and closed.

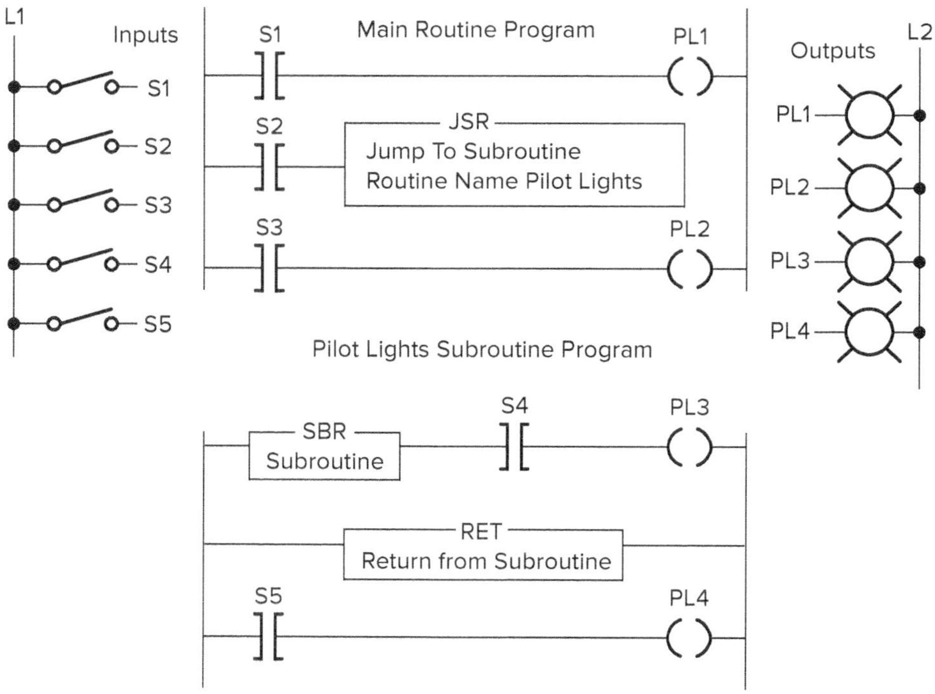

9-9 Construct and download the PLC program shown and perform each of the following force operations:

- Force PL1 On and note the affect, if any, on the operation of PL2.
- Force PL1 Off and note the affect, if any, on the operation of PL2.
- Force S2 On and note the affect, if any, on the operation of PL3.
- Force S2 Off and note the affect, if any, on the operation of PL3.

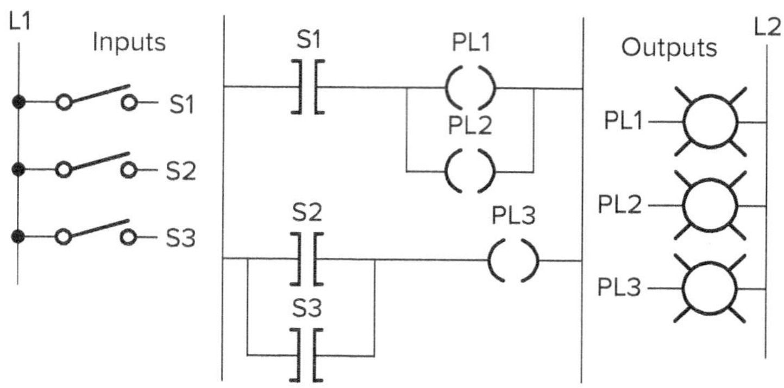

NAME ______________________________ DATE______________

9-10 Construct and download the MCR program shown. Operate the program according to the following sequence:

a) Close switches 1, 2, 3, 4, and 6, and allow timer T4:T2 to time out. What lights are on?

b) Open switch 1. What light is on now? Why did lights 1 and 2 go off?

c) Open switch 4 and close switch 5. Did light 1 go off? Why or why not?

d) What happened to the two timers when you disabled the MCR zone?

e) What happened to the two timers when you reenabled the MCR zone?

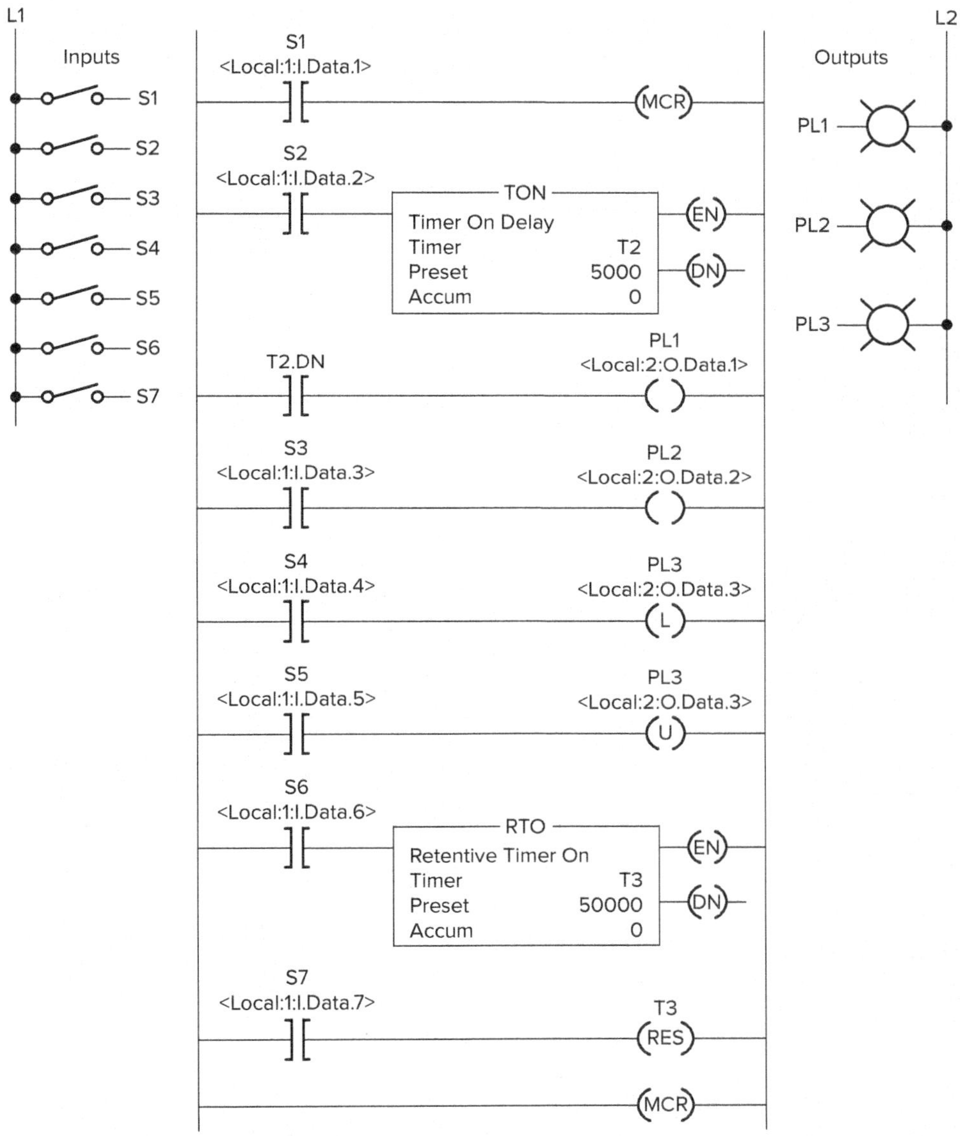

NAME ________________________________ DATE______________

9-11 Construct and download the Main and Subroutine programs shown. The purpose of the Subroutine is to find the average value of Variable_1 and Variable_2 and store the result in Sub_Result. This is accomplished by passing parameters to the subroutine and doing the math in the subroutine, and then returning the answer to the Main Routine through the RET instruction. Operate the program and observe the following steps:

- When the switch is closed, the data from Varable_1 and Variable_2 are copied into the ADD instruction.
- Here they are added together, and the result stored in Sub_Sum.
- This value is then divided by 2, and the result is stored in Sub_Result.
- The RET instruction then returns this average value to the JSR instruction of the Main Routine.

Main Routine

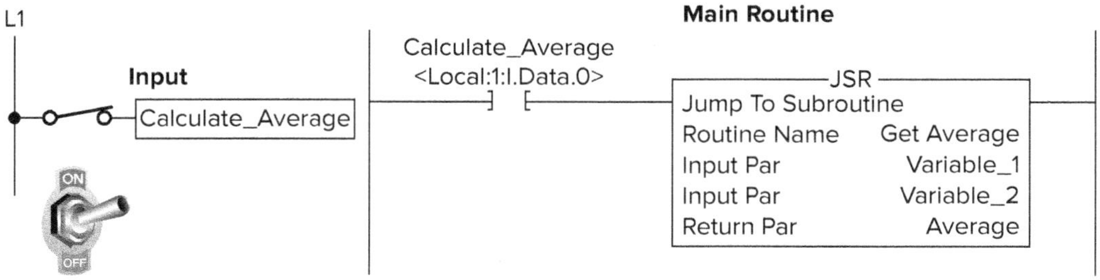

Calculate Average Of 2 Values Subroutine

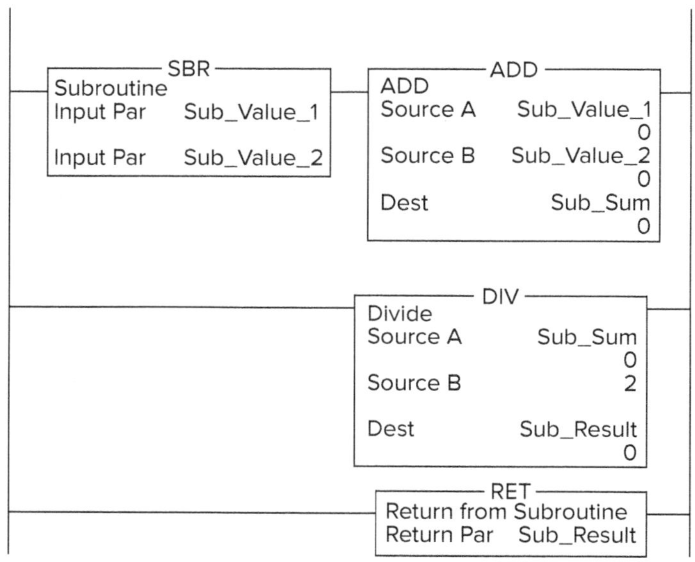

NAME ______________________________ DATE______________

CHAPTER 10

Data Manipulation Instructions

10-1 Construct and download the PLC Move (MOV) instruction program shown. Test your program by viewing its operation with the Tag Monitor window.

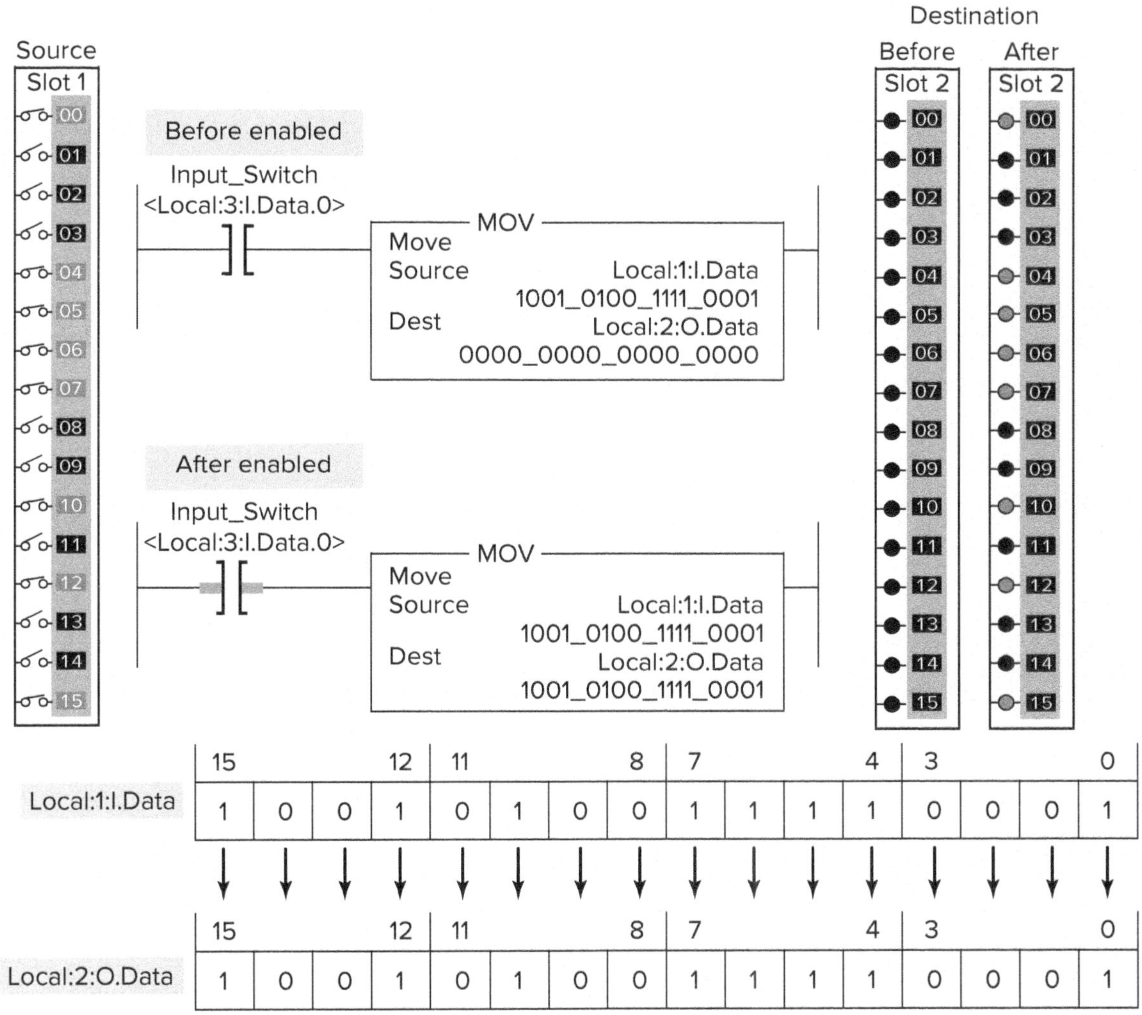

NAME ________________________________ DATE______________

10-2 Construct and download the PLC Masked Move (MVM) instruction program shown. Test your program by viewing its operation with the Tag Monitor window.

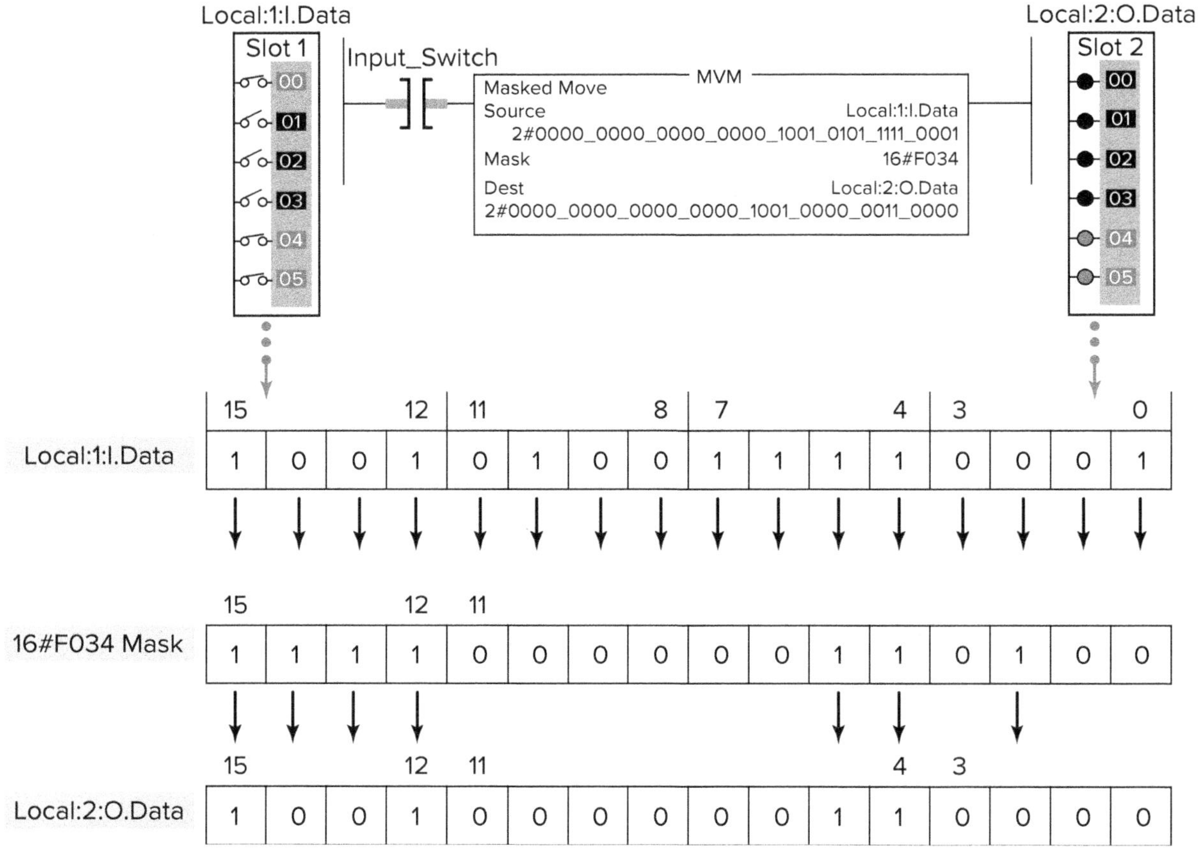

NAME ______________________ DATE ____________

10-3 Construct and download the Bit Field Distribution (BTD) instruction program shown. Test your program by viewing its operation with the Tag Monitor window.

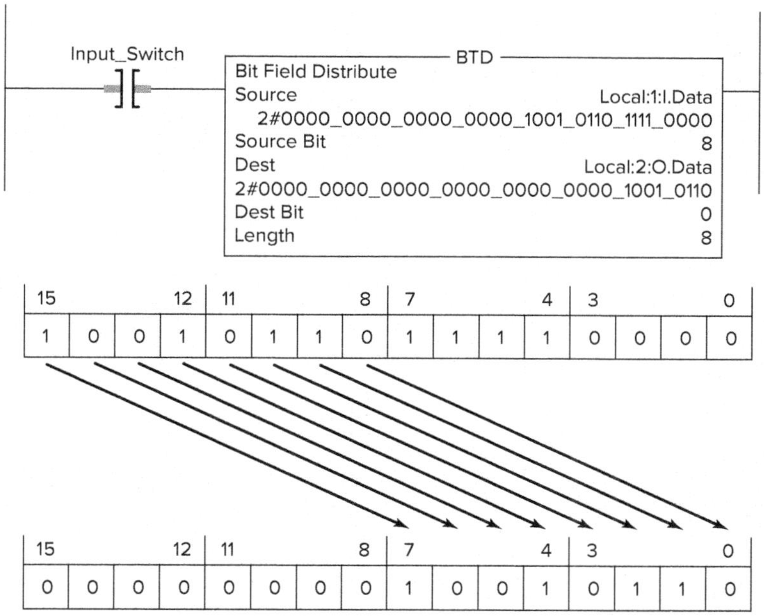

10-4(a) Construct and download the PLC move instruction timer preset program shown. Test your program for proper operation.

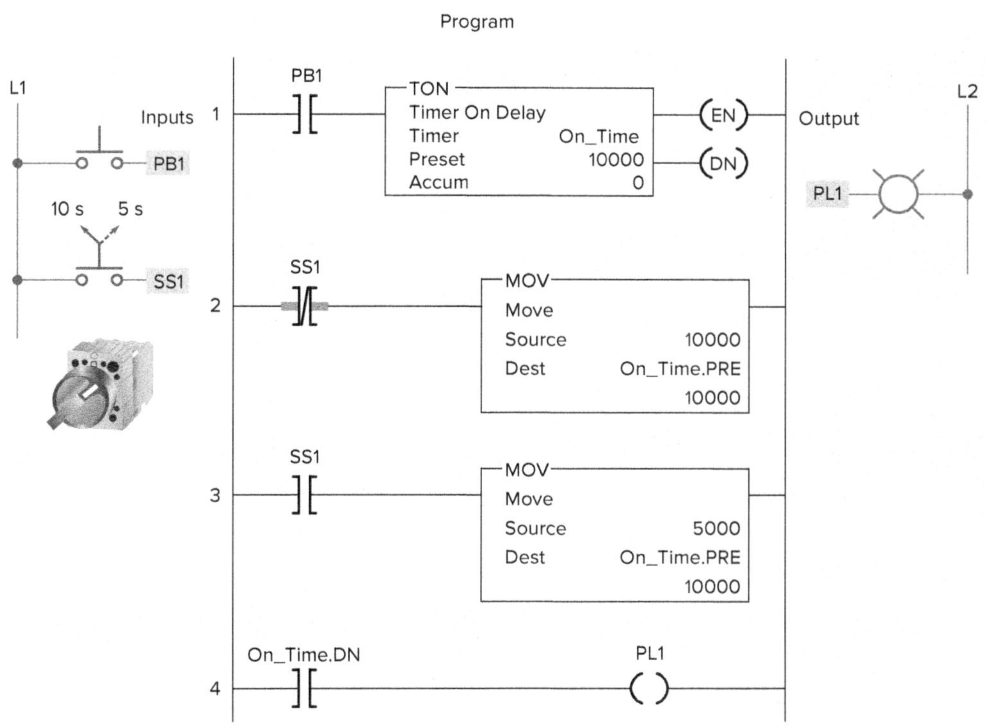

NAME ______________________________ DATE______________

10-4(b) Modify the program for time settings for 10 and 25 seconds and add 2 pilot lights that indicate which setting is being selected.

10-5(a) Construct and download the PLC move instruction counter preset program shown. Test your program for proper operation.

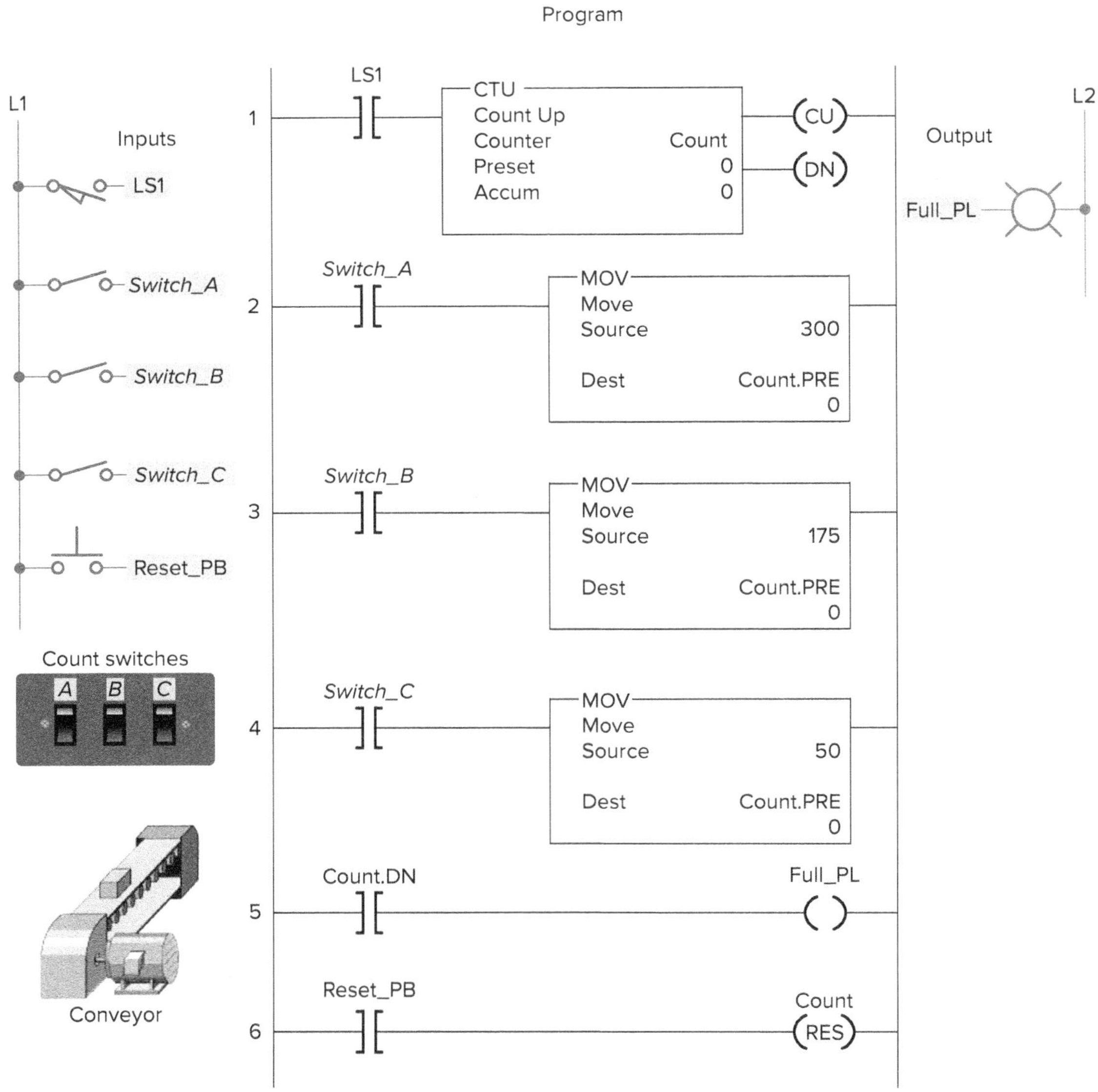

NAME ______________________________ DATE______________

10-5(b) Modify the program for count settings of 5, 10, 15 and add 3 pilot lights that indicate which setting is being selected.

10-6 Construct and download the array Copy (COP) instruction program shown. Test your program by inserting data into the Source and confirming its transfer to the Destination in the Tag Monitor window.

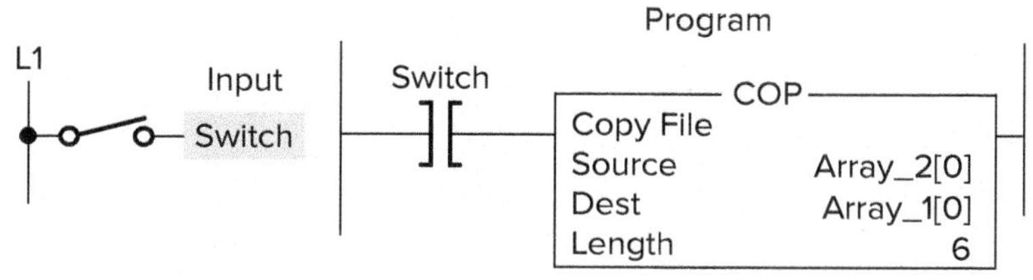

10-7 Construct and download the drink dispenser program shown. Test your program by inserting data into the Source and confirming its transfer of data to the Destination in the Tag Monitor window.

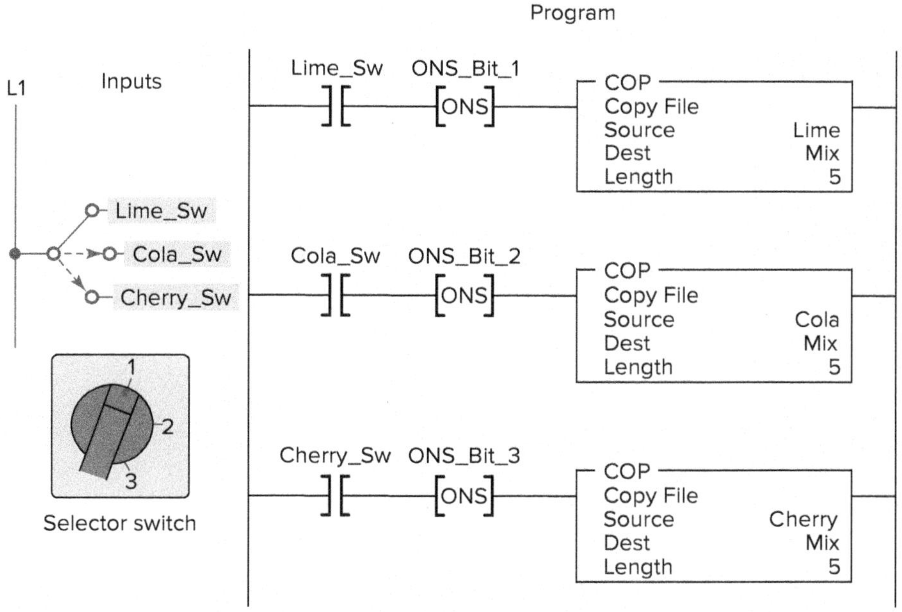

NAME ____________________ DATE ______________

10-8 Construct and download the Equal (EQU) program shown. Test your program for proper operation.

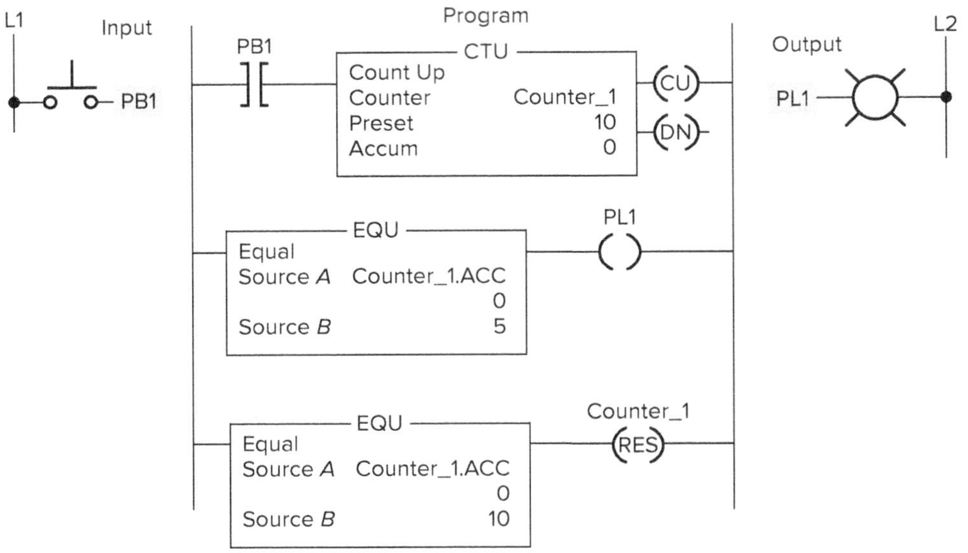

10-9 Construct and download the Not Equal (NEQ) program shown. Test your program for proper operation.

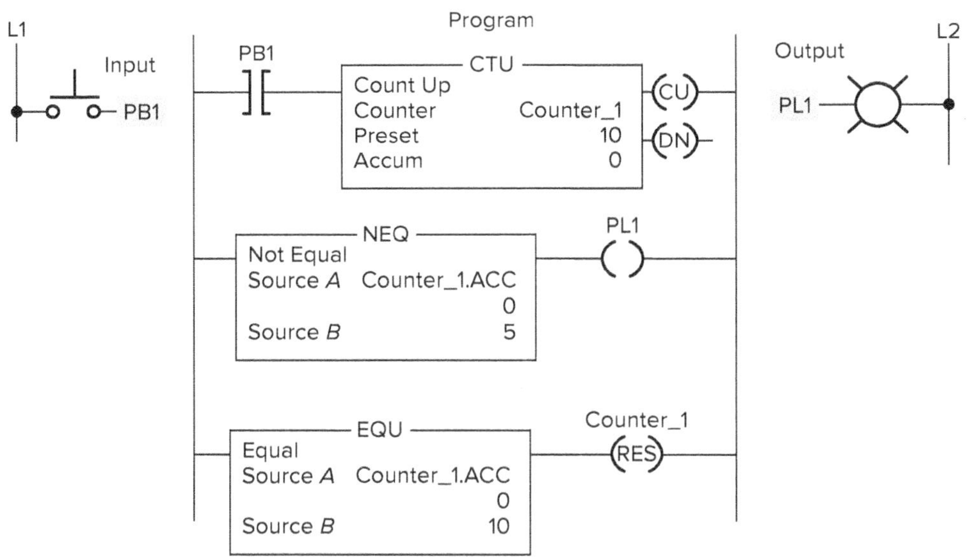

NAME ______________________ DATE ____________

10-10 Construct and download the Greater Than (GRT) program shown. Test your program for proper operation.

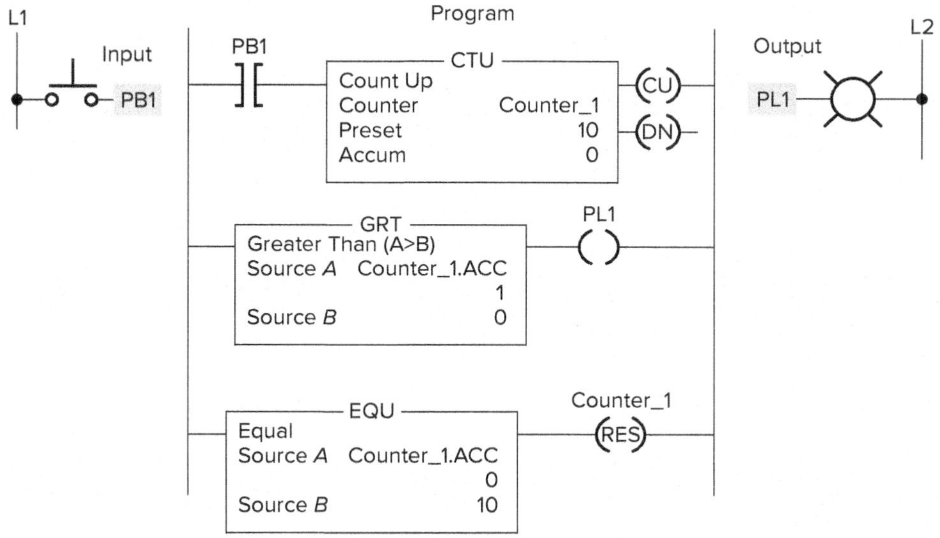

10-11 Construct and download the Less Than (LES) program shown. Test your program for proper operation.

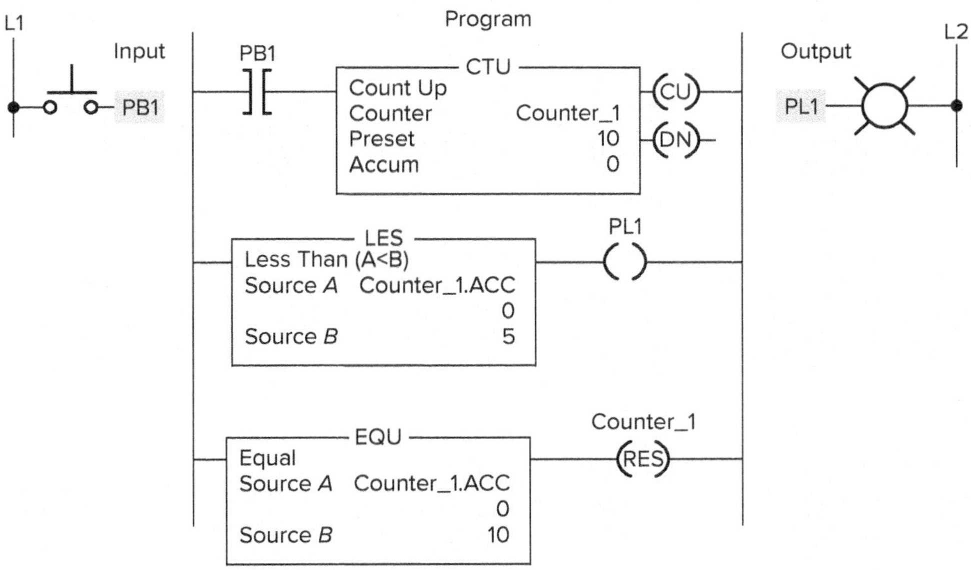

NAME ______________________________ DATE______________

10-12 Construct and download the Greater Than or Equal (GEQ) program shown. Test your program for proper operation.

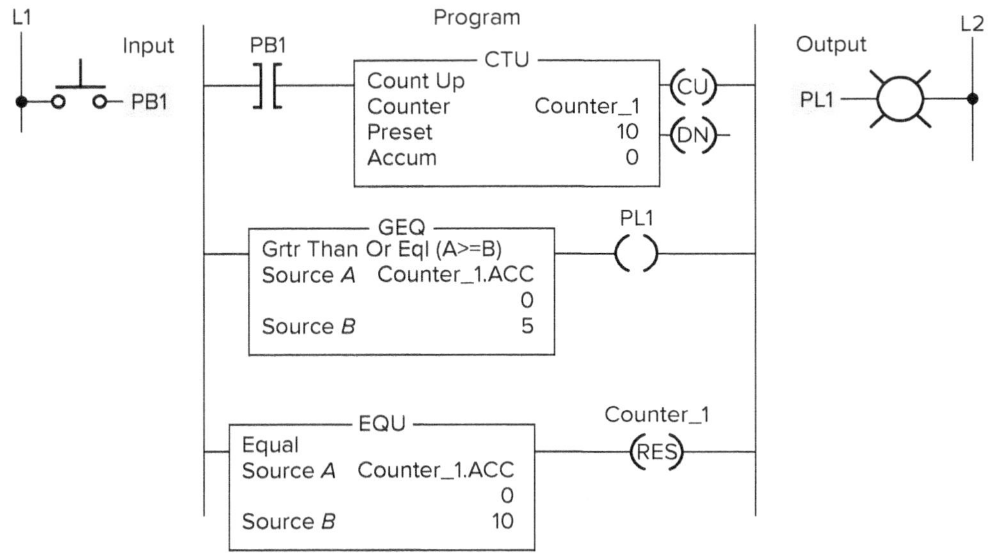

10-13 Construct and download the Less Than or Equal (LEQ) program shown. Test your program for proper operation.

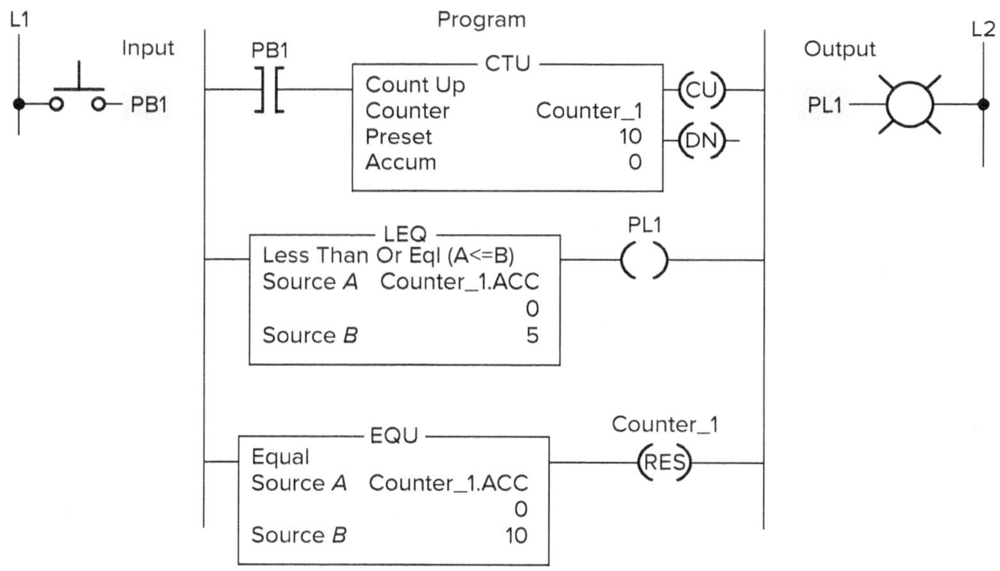

NAME ____________________ DATE ____________

10-14 Construct and download the counter triggering of outputs program shown. Test your program for proper operation.

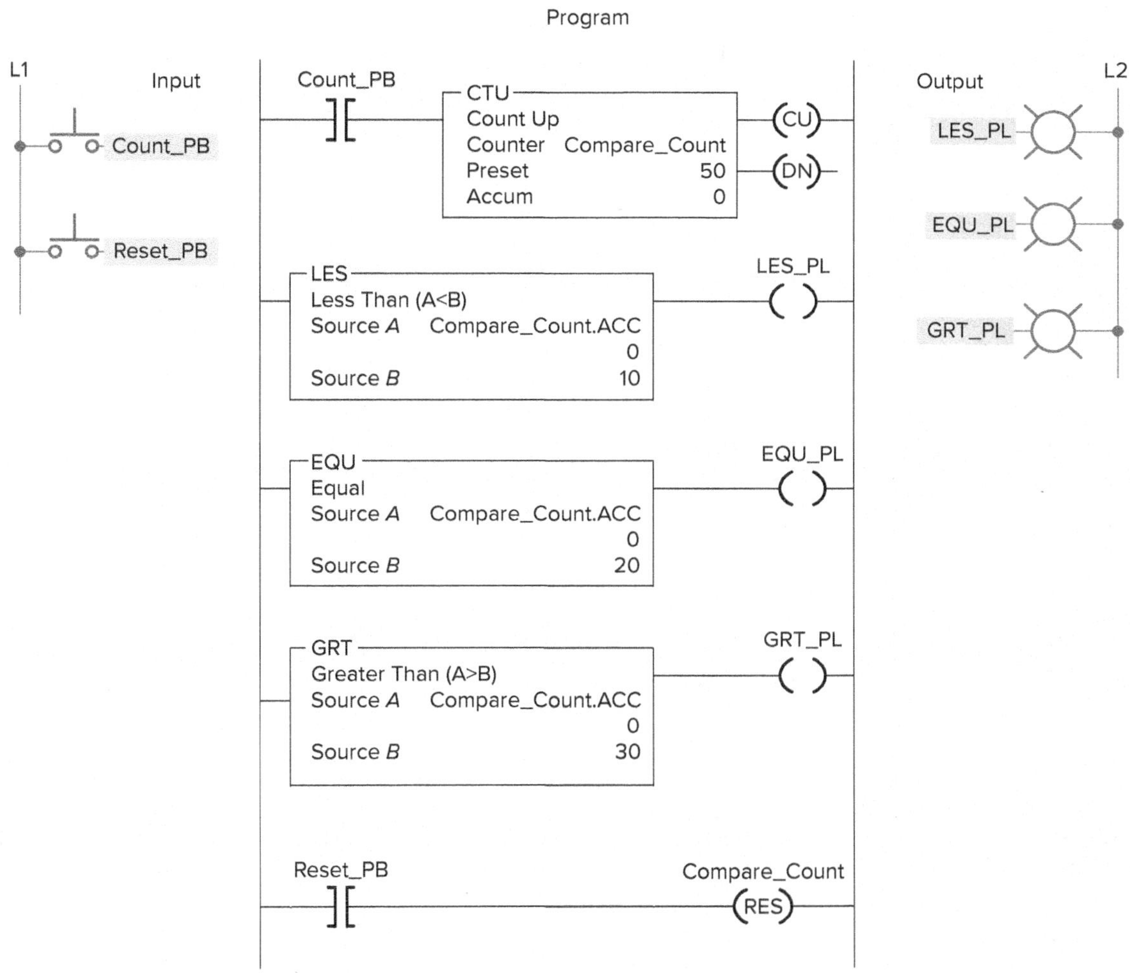

NAME ______________________________ DATE______________

10-15 Construct and download the Limit (LIM) instruction program shown. Test your program for proper operation.

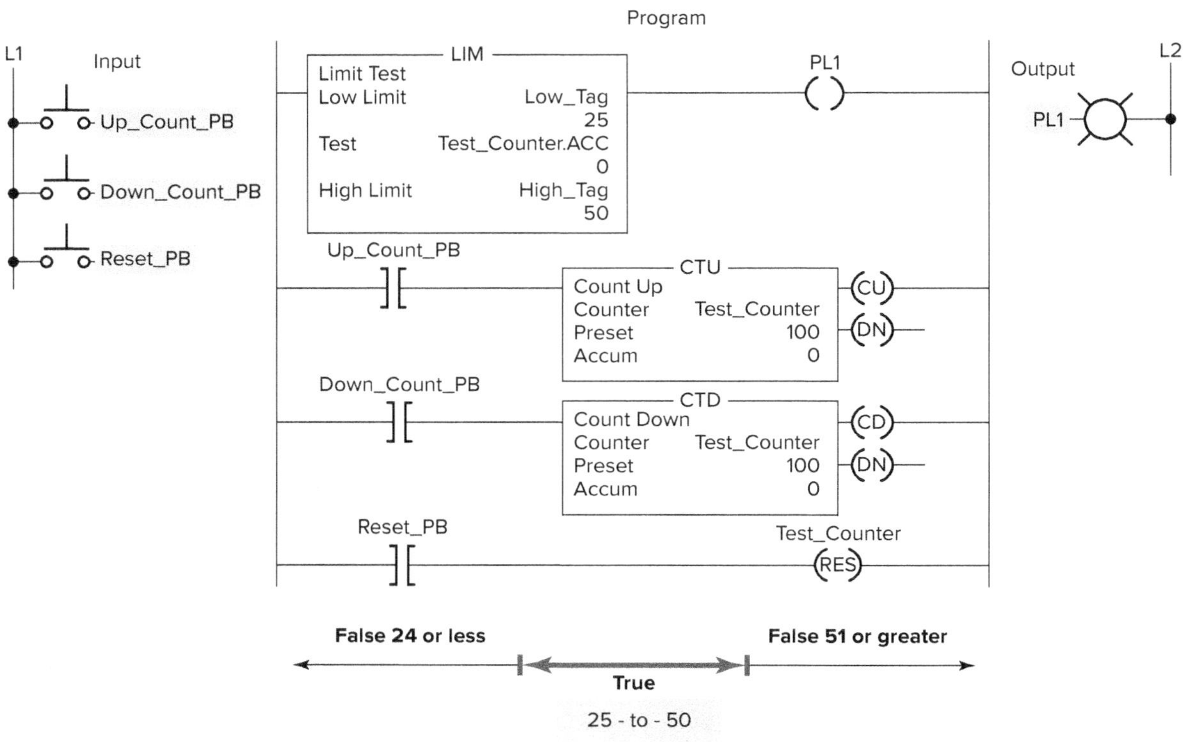

10-16 Construct and download the PLC limit instruction used in the timer monitoring program shown. Test your program for proper operation.

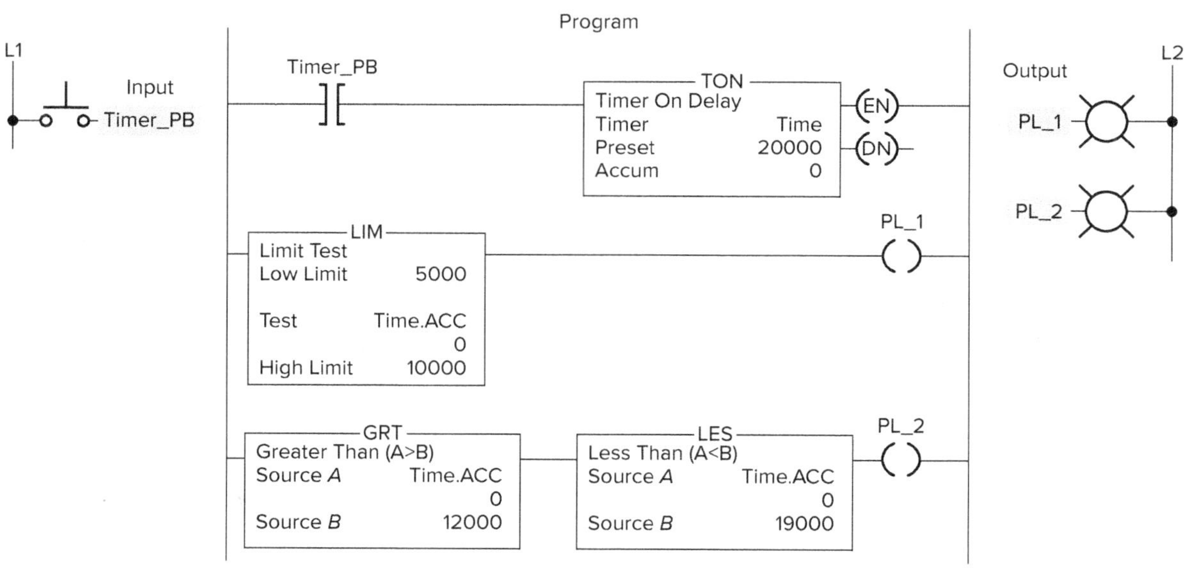

NAME ______________________________ DATE______________

10-17 Construct and download the GEQ instruction used in the solenoid valve control program shown. Test your program for proper operation.

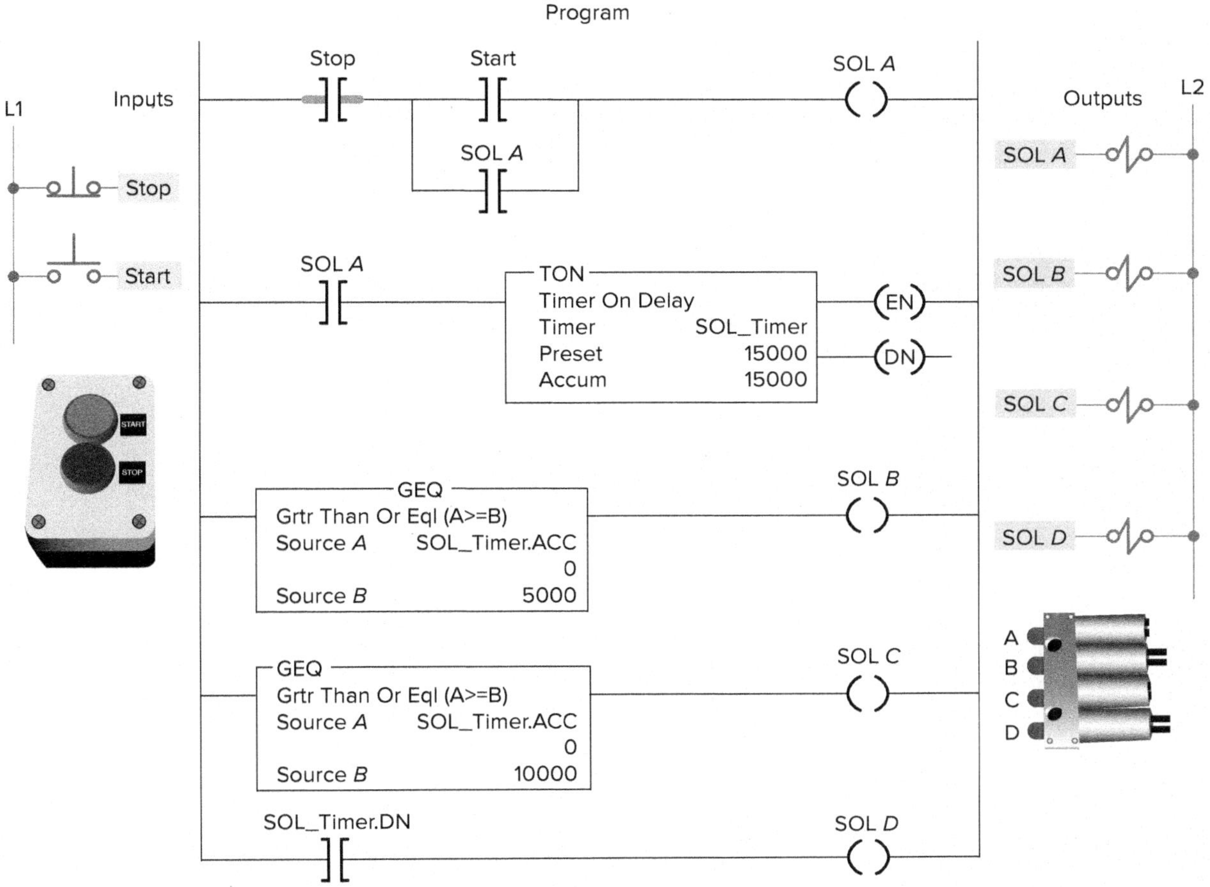

NAME ________________________________ DATE______________

10-18 Construct and download the on-delay timer program implemented in conjunction with the equal (EQU) instruction. Test your program for proper operation.

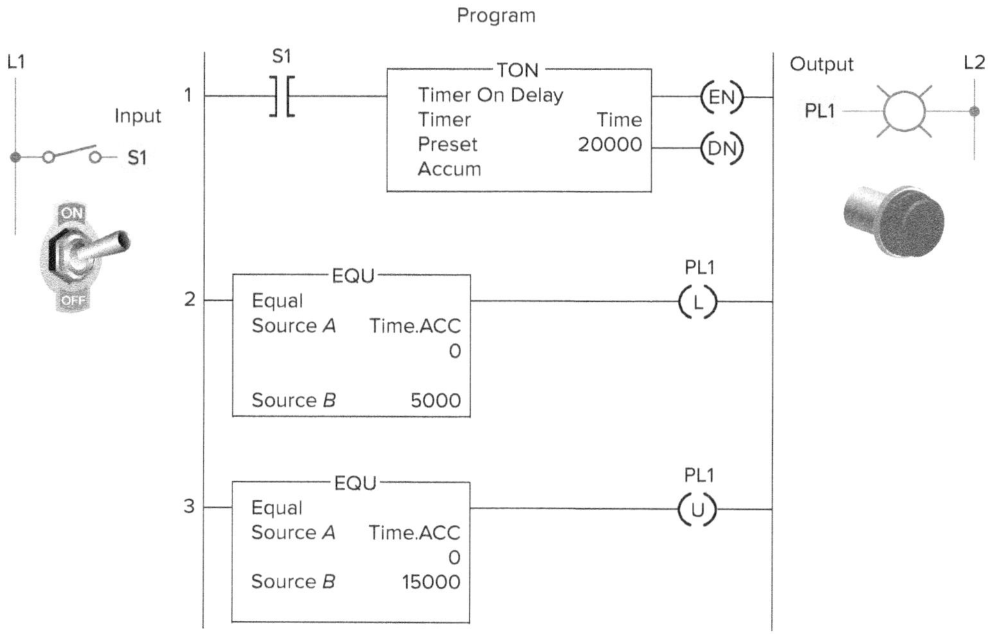

10-19 Construct and download the up-counter program implemented using the LES instruction. Test your program for proper operation.

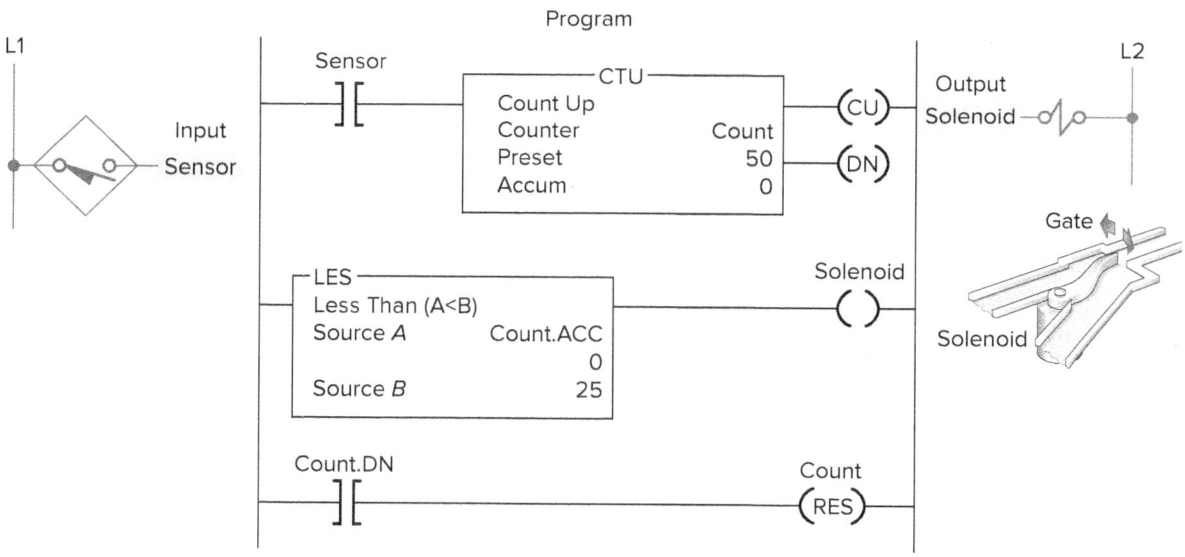

NAME ______________________________ DATE ______________

10-20(a) Construct and download the program for monitoring the setting of a thumbwheel switch shown. Test your program for proper operation.

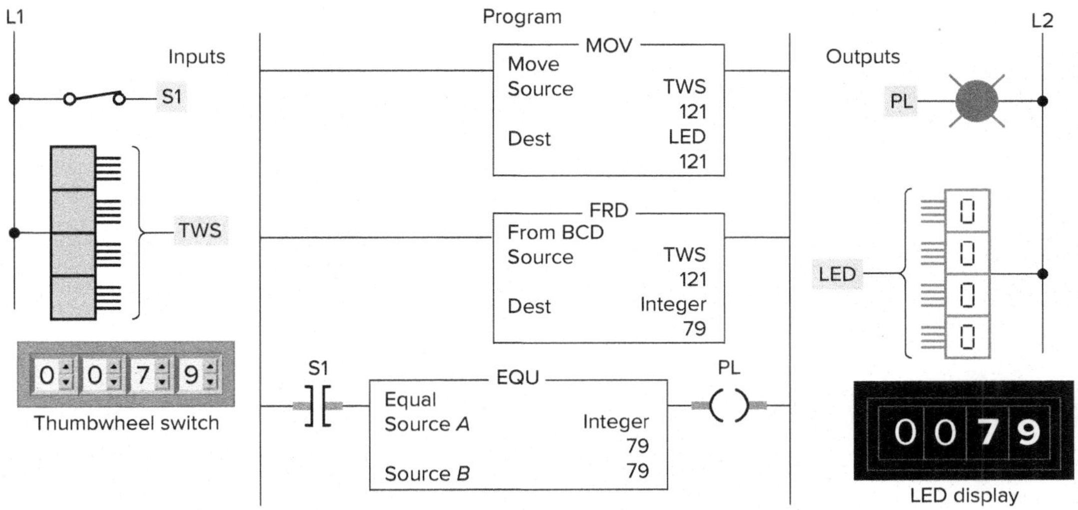

10-20(b) Modify your program by adding 2 more pilot lights, one to indicate a setting that is less than 79 and the other to indicate the setting is greater than 79.

NAME ______________________________ DATE______________

10-21(a) Construct and download the set-point control program shown. Test your program for proper operation.

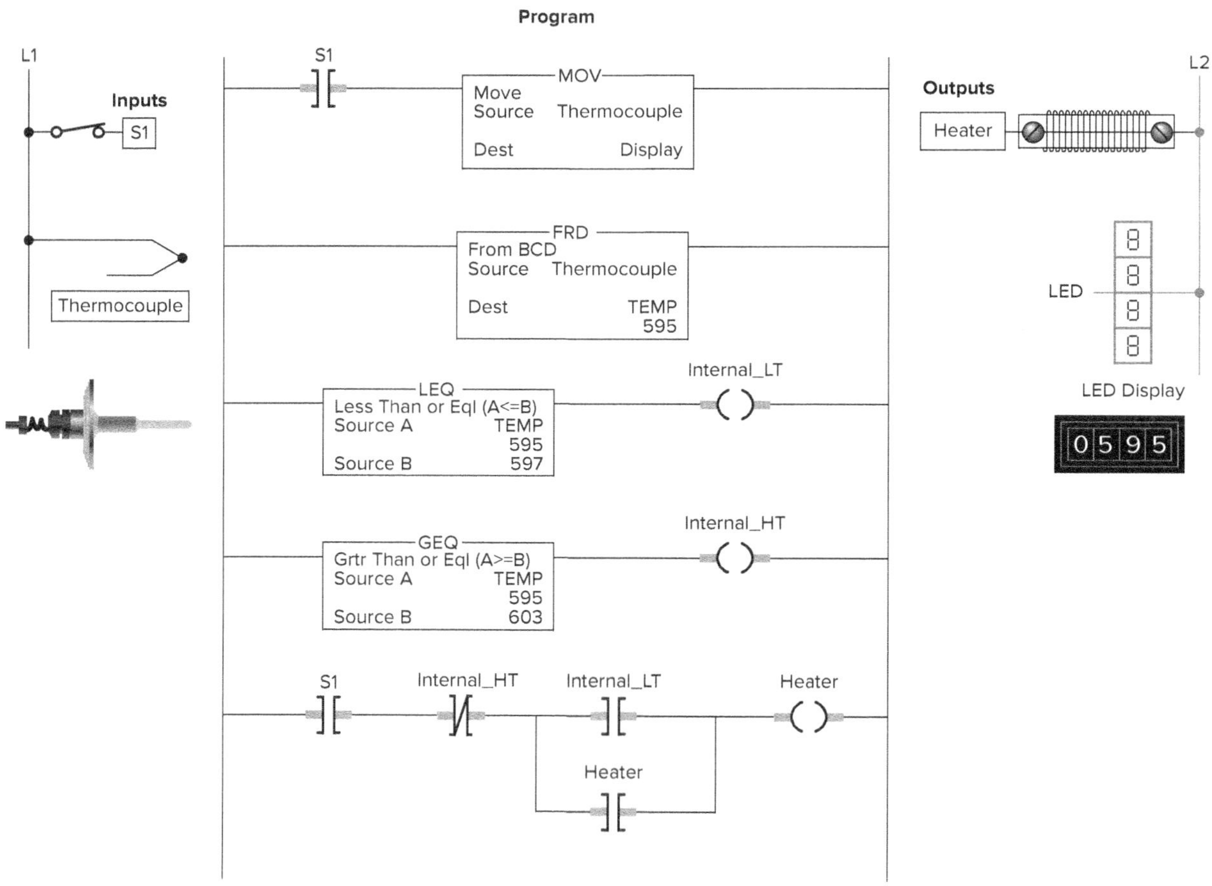

10-21(b) Modify the program to include alarm lights that come on if the temperature rises above 610°F or drops below 590°F. Once on, each alarm light stays on until manually reset with a pushbutton.

NAME ______________________________ DATE______________

10-22 The data compare program shown uses a thumbwheel switch interface module for the changing variable. Construct and download the program. Test your program for proper operation.

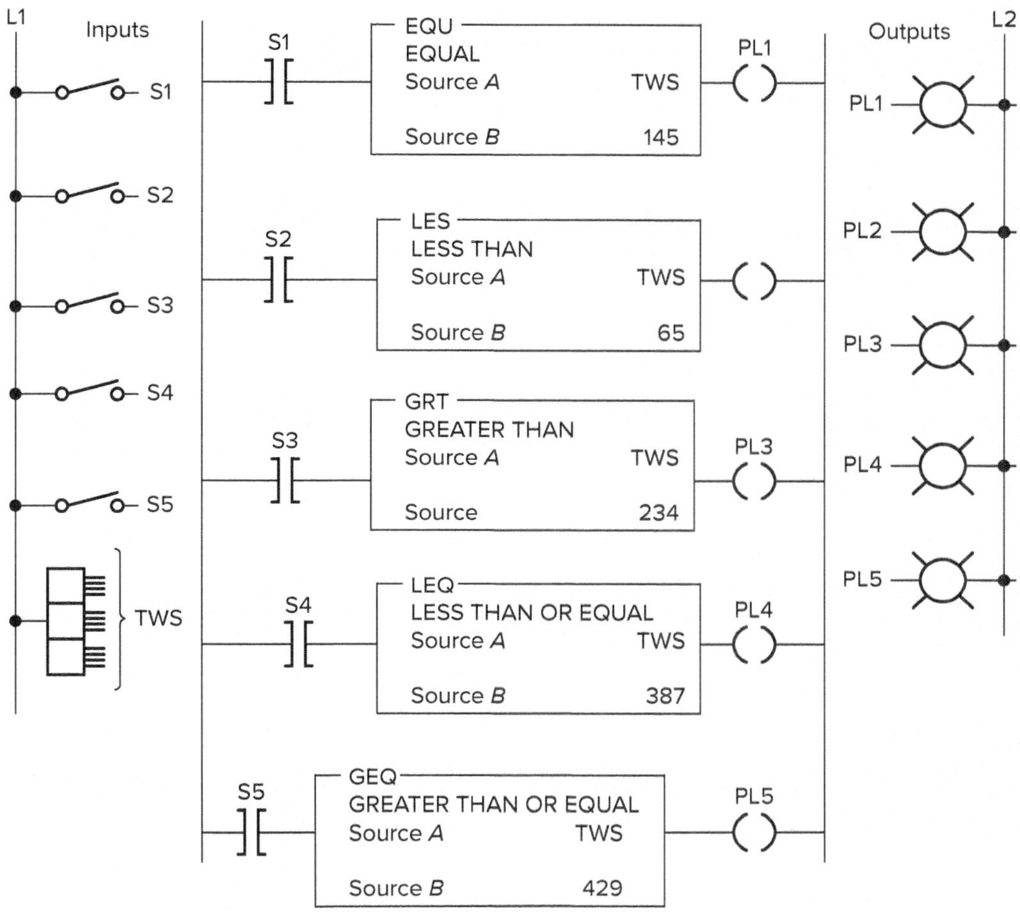

NAME ______________________ DATE______________

CHAPTER 11

Math Instructions

11-1 Construct the ADD instruction program shown. Download your program to the PLC and test its operation.

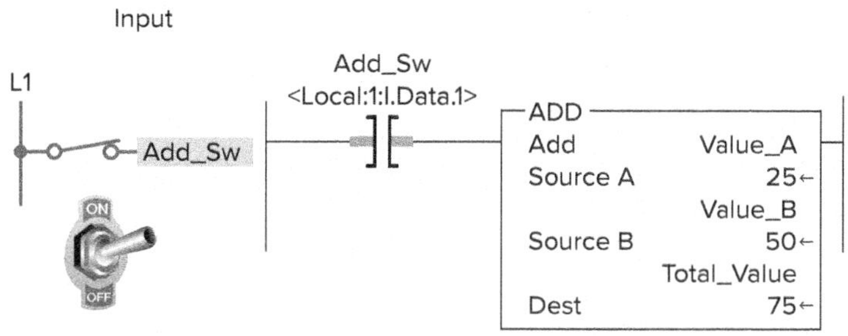

Tag Name	Value	Style	Data Type
[+]-Total_Value	75	Decimal	DINT
[+]-Value_A	25	Decimal	DINT
[+]-Value_B	50	Decimal	DINT
Add_Sw	1	Decimal	BOOL

11-2 Construct the SUB instruction program shown. Download your program to the PLC and test its operation.

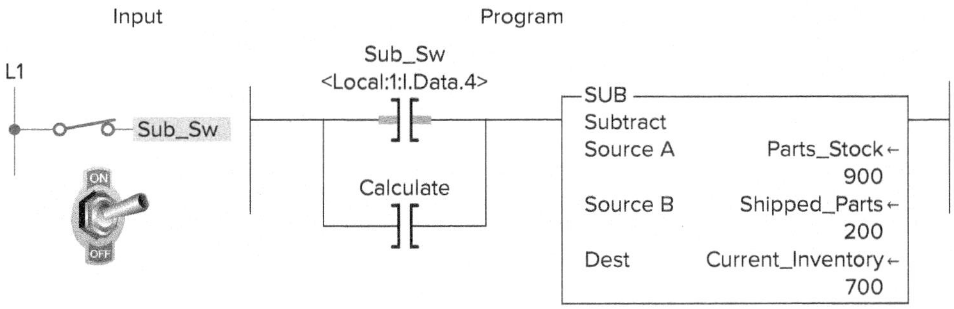

Tag Name	Value	Style	Data Type
[+]-Parts_Stock	900	Decimal	DINT
[+]-Shipped_Parts	200	Decimal	DINT
[+]-Current_Inventory	700	Decimal	DINT
Sub_Sw	1	Decimal	BOOL
Calculate	0	Decimal	BOOL

NAME ____________________ DATE ____________

11-3 Construct the MUL instruction program shown. Download your program to the PLC and test its operation.

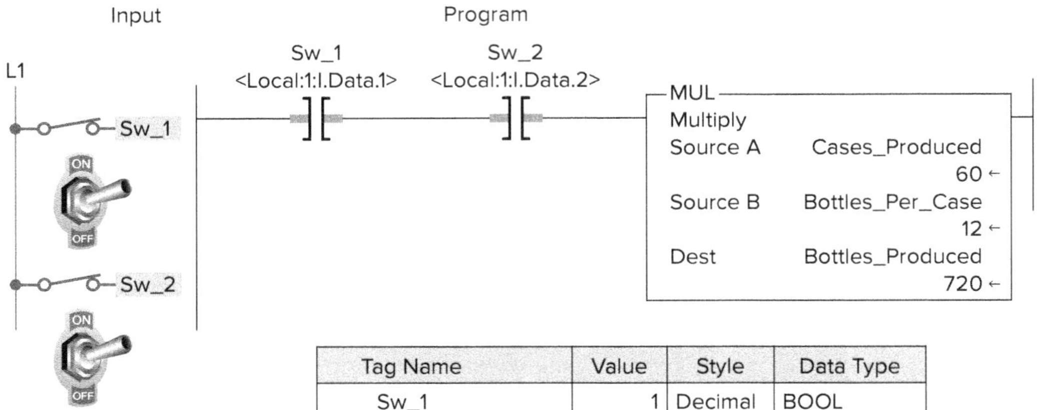

Tag Name	Value	Style	Data Type
Sw_1	1	Decimal	BOOL
Sw_2	1	Decimal	BOOL
[+]-Cases_Produced	60	Decimal	DINT
[+]-Bottles_Per_Case	12	Decimal	DINT
[+]-Bottles_Produced	720	Decimal	DINT

11-4 Construct the DIV instruction program shown. Download your program to the PLC and test its operation.

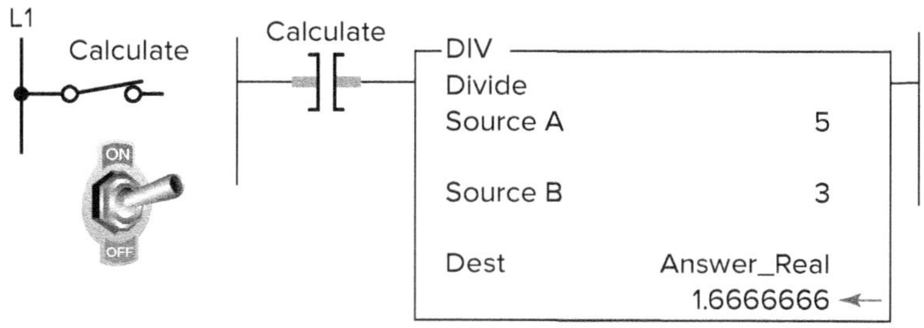

Tag Name	Value	Style	Data Type
Calculate	1	Decimal	BOOL
Answer_Real	1.6666666	Float	REAL

NAME ______________________________ DATE ______________

11-5 Construct and download the PLC ADD instruction program used as part of a three-conveyor parts tracking system. Test your program for proper operation.

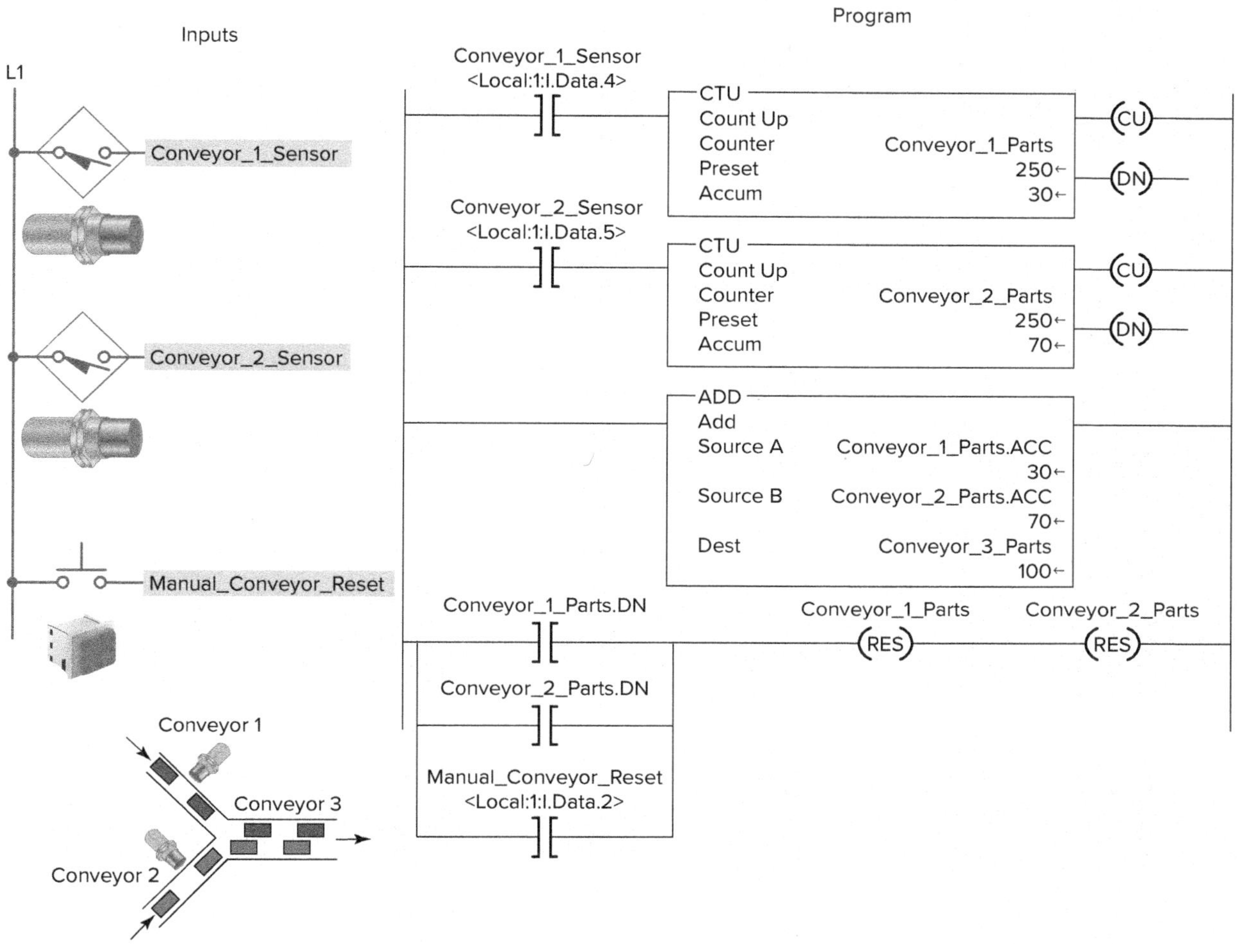

NAME ______________________________ DATE______________

11-6 Construct and download the vessel overfill alarm program shown. Test your program for proper operation.

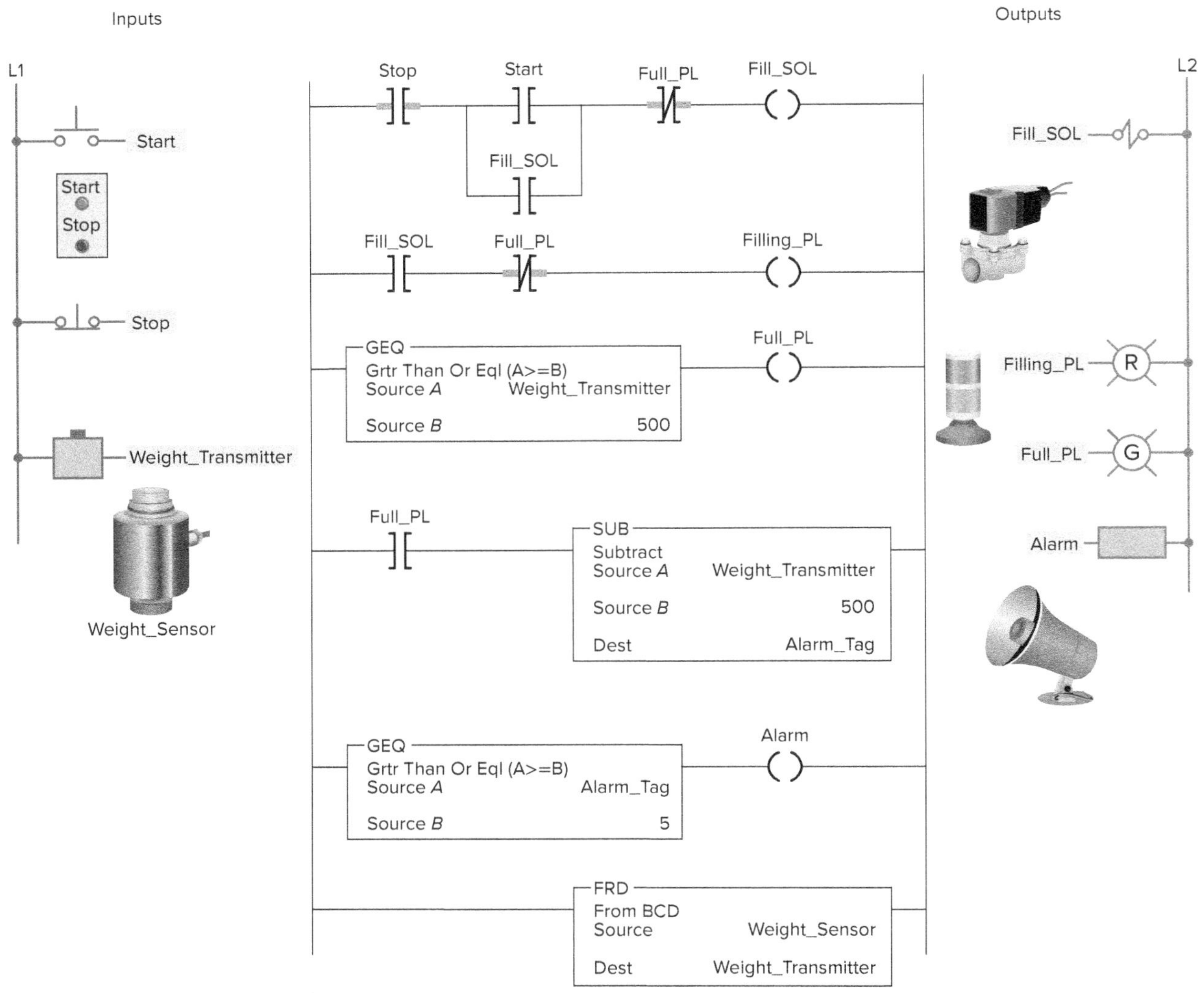

NAME ______________________________ DATE______________

11-7 Construct and download the oven temperature control program shown. Test your program for proper operation.

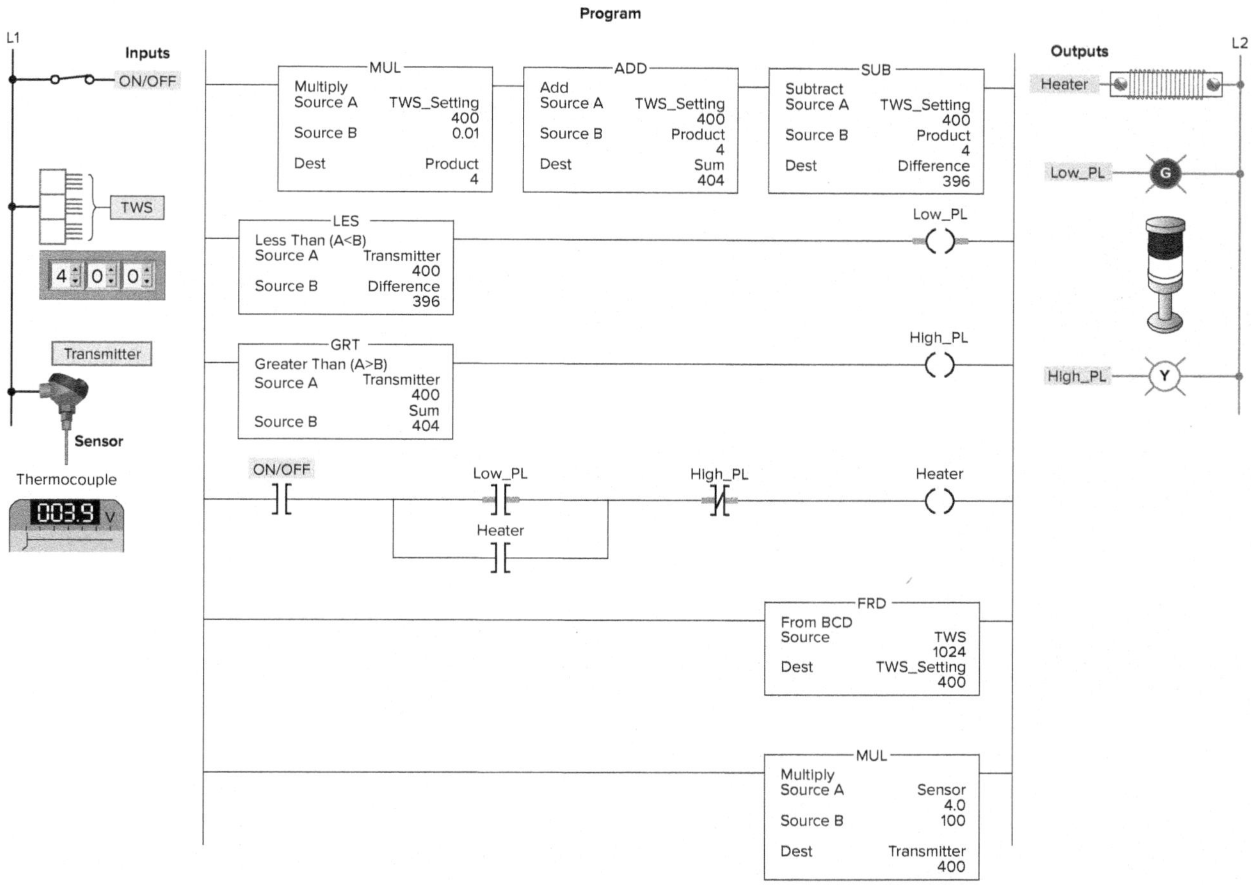

11-8 Construct and download the program used to convert Celsius temperature to Fahrenheit shown. Test your program for proper operation.

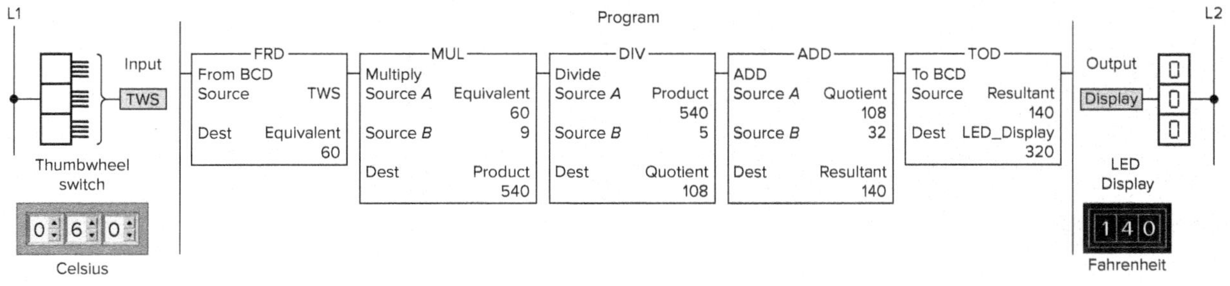

NAME ______________________________ DATE______________

11-9 Construct and download the cooling tower fan program shown. Test your program for proper operation.

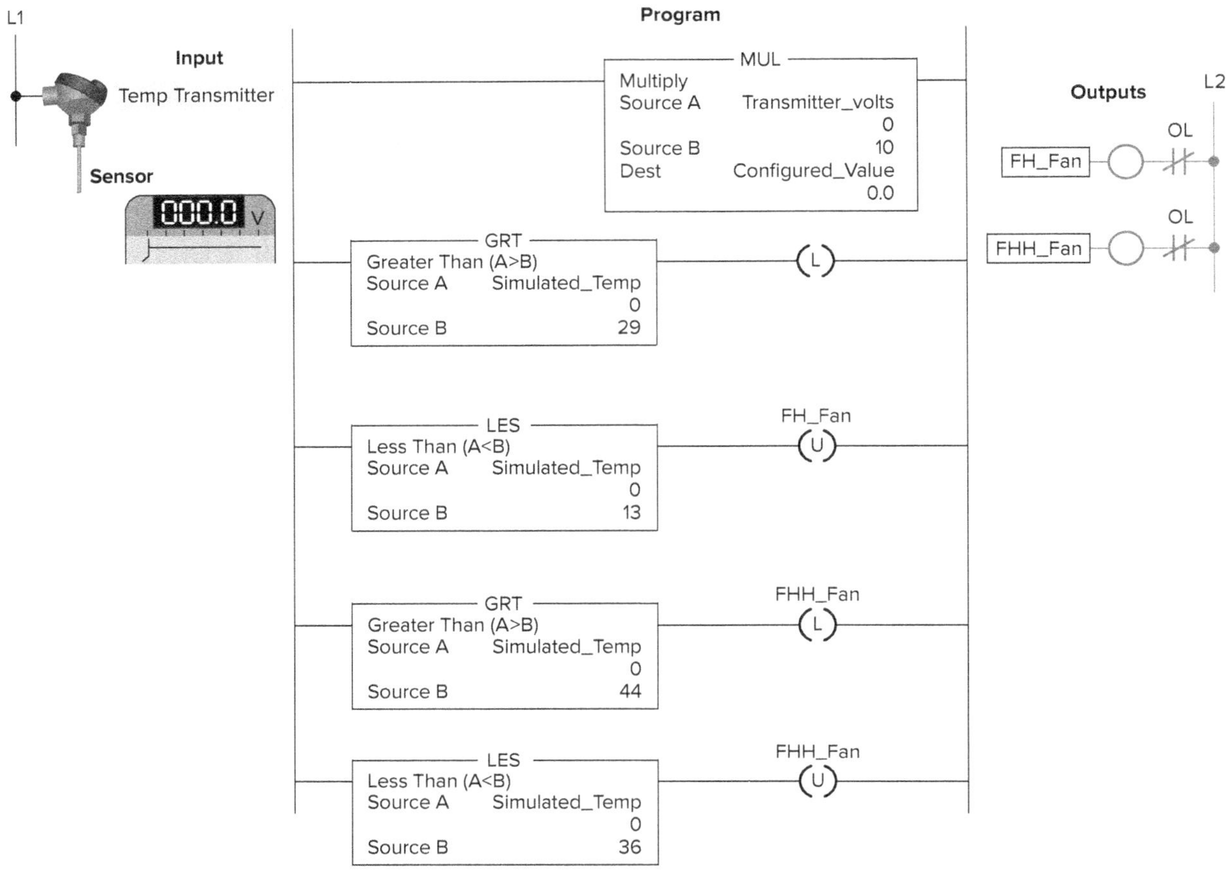

NAME ______________________________ DATE______________

11-10 Construct and download the program used to display the lapsed time period of a timer shown. Test your program for proper operation.

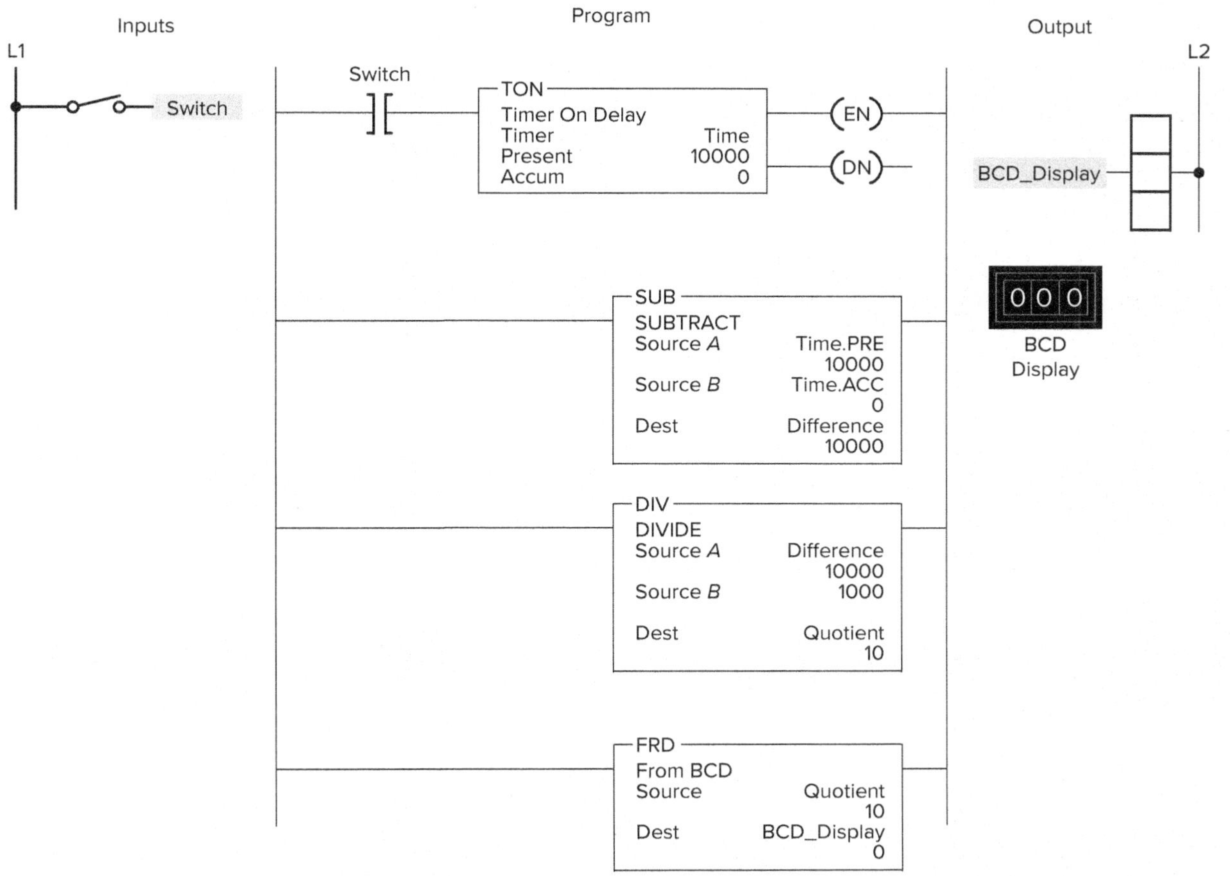

11-11 Construct the SQR (Square Root) instruction program shown. Download your program to the PLC and test its operation.

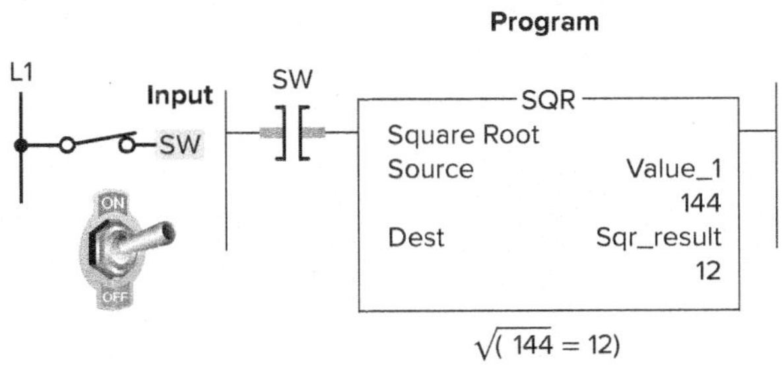

NAME ________________________________ DATE______________

11-12 Construct the NEG (Negate) instruction program shown. Download your program to the PLC and test its operation.

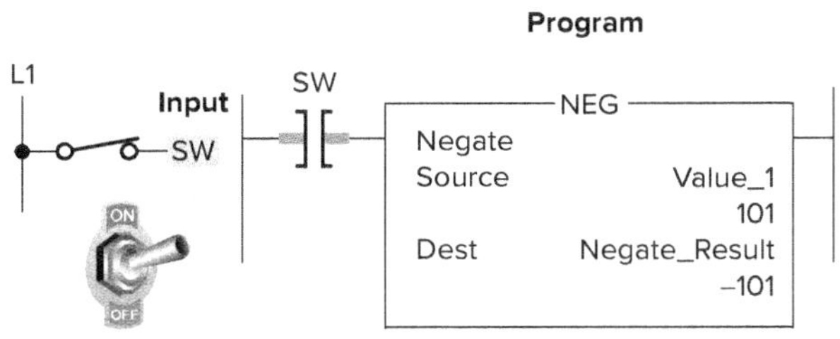

11-13 Construct the CLR (Clear) instruction program shown. Download your program to the PLC and test its operation.

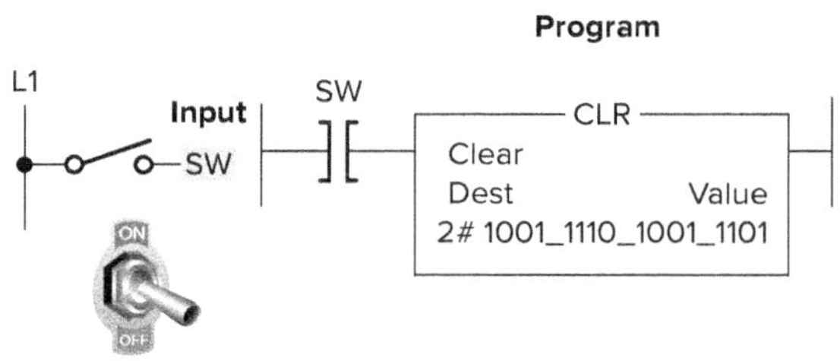

11-14 Construct the TOD (convert to BCD) instruction program shown. Download your program to the PLC and test its operation.

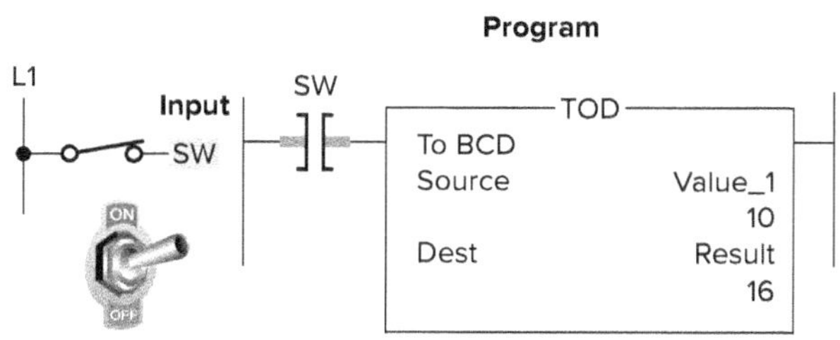

11-15 Construct the FRD (convert from BCD) instruction program shown. Download your program to the PLC and test its operation.

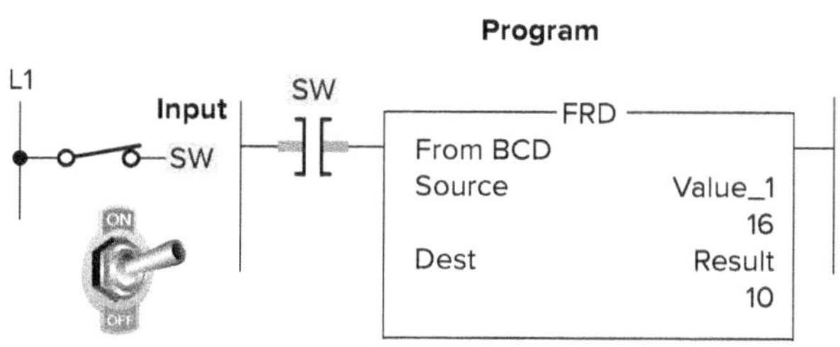

NAME ______________________________ DATE______________

11-16 The program shown is a voltage ramp application executed by configuring the analog output module as illustrated in the text. Construct and download your program to the PLC and test its operation.

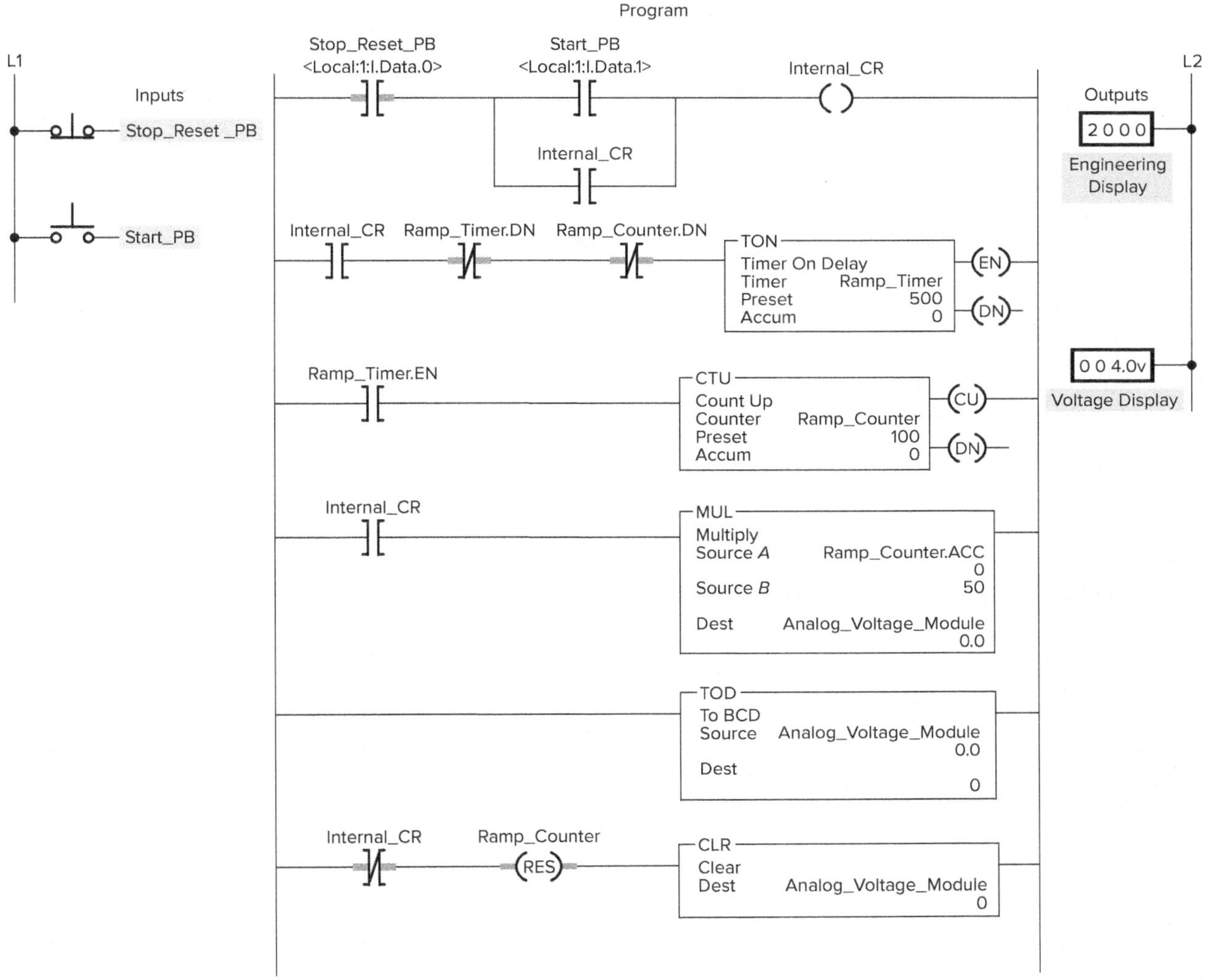

11-17 The program shown illustrates two ways to accomplish output scaling of the previous program. The program is executed as follows:

- The incrementing value from the counter gets multiplied by 50 and sent to the Slot 6 BCD Display as well as to channel 0 of the Slot 8, four channel analog output module.
- The channel 0 module card is setup to scale an engineering range of 0 to 5000 down to an output voltage range of 0.0 to 10.0 V.

NAME ________________________________ DATE______________

- In addition, the Raw Output Data is also used with a DIV instruction to scale the value down to the appropriate output range by dividing the Raw Output Data value by 500, resulting in a value from 0 to 10.

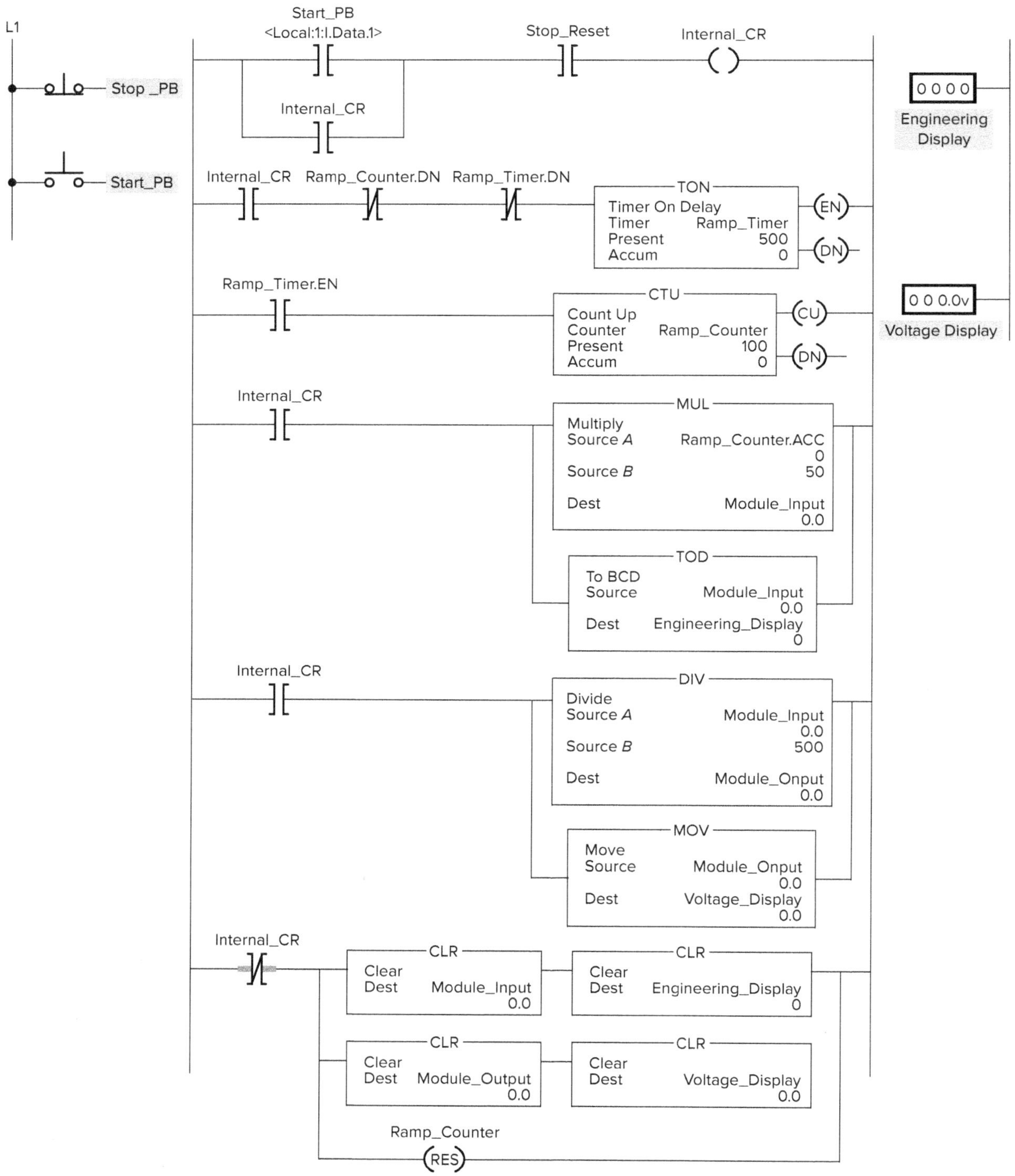

NAME ___________________________________ DATE_______________

11-18 Three conveyors are delivering the same parts in different packages. A package can hold 12, 24, or 18 parts. Proximity switches installed on each of the conveyor lines are used to advance the accumulated value of the three counters. Write a program that uses multiply and add instructions to calculate the sum of the parts and display this number as a BCD format LED display. Download your program to the PLC and test its operation.

11-19 Two conveyor lines are outputting two identical products. Write a program using two sensors and counters along with math instructions which will continuously monitor the sum and average of products being outputted. Download your program to the PLC and test its operation.

11-20 Two parts conveyor lines, A and B, feed a main conveyor line M. A third conveyor line, R, removes rejected parts a short distance away from the main conveyor. Conveyors A, B, and R have parts counters connected to them. Construct a PLC program to obtain the parts output of main conveyor M and display this number on a four-digit BCD format LED display.

CHAPTER 12

Sequencer and Shift Register Instructions

12-1 Construct and download the manually operated, traffic control program. Test your program for proper operation.

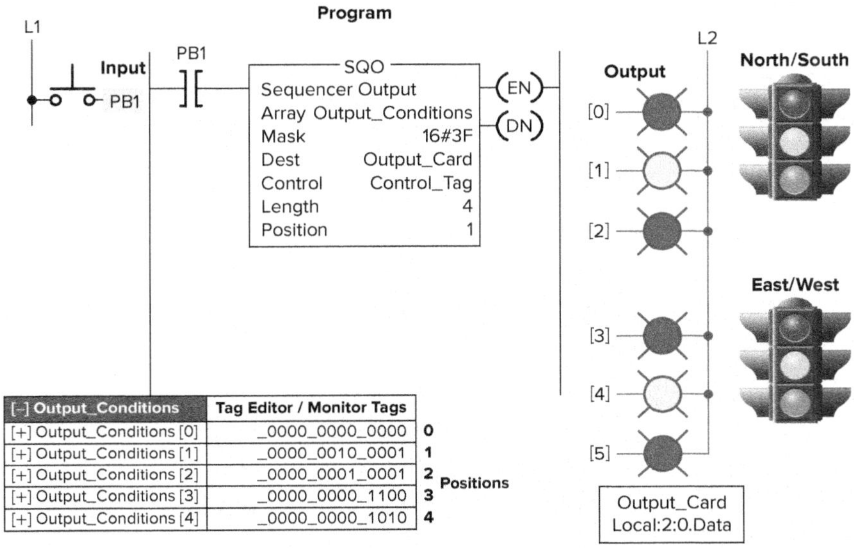

[-] Output_Conditions	Tag Editor / Monitor Tags	
[+] Output_Conditions [0]	_0000_0000_0000	0
[+] Output_Conditions [1]	_0000_0010_0001	1
[+] Output_Conditions [2]	_0000_0001_0001	2
[+] Output_Conditions [3]	_0000_0000_1100	3
[+] Output_Conditions [4]	_0000_0000_1010	4

12-2 Construct and download the sequencer mask move program shown. Test your program for proper operation.

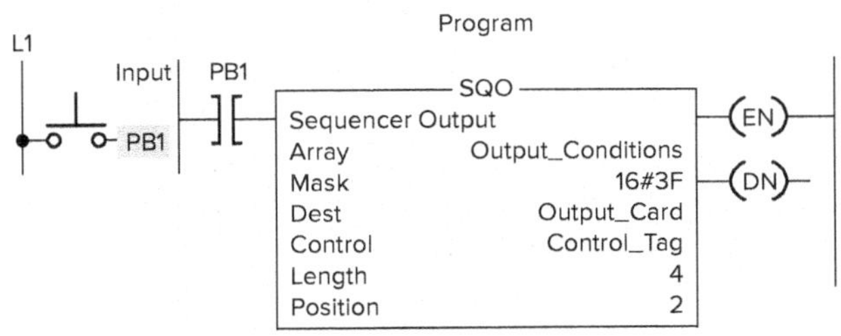

	15	14	13	12	11	10	9	8	7	6	5	4	3	2	1	0	
Array Output_Conditions	0	0	0	0	0	0	0	0	0	0	0	0	0	0	0	0	Step 0
	0	0	0	0	0	0	0	0	0	0	1	0	0	0	0	1	Step 1
	0	0	0	0	0	0	0	0	0	0	0	1	0	0	0	1	Step 2
	0	0	0	0	0	0	0	0	0	0	0	0	1	1	0	0	Step 3
	0	0	0	0	0	0	0	0	0	0	0	0	1	0	1	0	Step 4
16#3F Mask	0	0	0	0	0	0	0	0	0	0	1	1	1	1	1	1	
Output_Card Local:2:O.Data Destination	0	0	1	0	0	1	1	0	1	0	0	1	0	0	0	1	

NAME ______________________________ DATE ______________

12-3 Construct and download the time-driven sequencer output program shown. Test your program for proper operation.

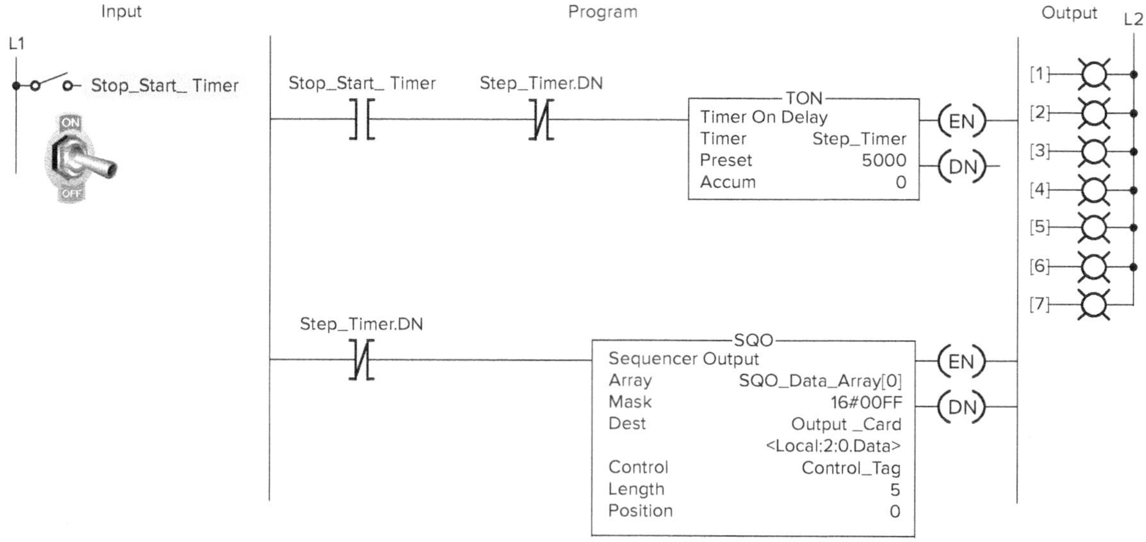

Output states entered into sequencer data array

[-] SQO_Data_Array	
[+] SQO_Data_Array[0]	2#0000_0000_0000
[+] SQO_Data_Array[1]	2#0000_0000_0010
[+] SQO_Data_Array[2]	2#0000_0000_0110
[+] SQO_Data_Array[3]	2#0000_0001_0000
[+] SQO_Data_Array[4]	2#0000_1100_1000
[+] SQO_Data_Array[5]	2#0000_1111_0000

Outputs which are On for each step

Outputs								Step
7	6	5	4	3	2	1	0	
								0
						On		1
					On	On		2
			On					3
On	On			On				4
On	On	On	On					5

12-4 Design a PLC program that will emulate the sequencer chart shown. Test your program for proper operation.

Data bits

Position	7	6	5	4	3	2	1	0
1	*							*
2		*					*	
3			*			*		
4				*	*			
5	*	*	*			*	*	*
6			*			*		
7		*					*	
8	*							*

* Indicates that output is energized

Matrix-style chart

Output card
0000000010000001
0000000001000010
0000000000100100
0000000000011000
0000000011100111
0000000000100100
0000000001000010
0000000010000001

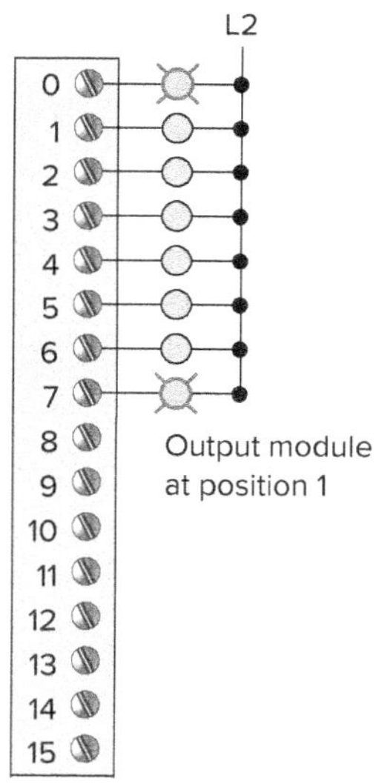

Output module at position 1

NAME ______________________________ DATE______________

12-5 Construct and download the automatically operated traffic control program shown. Test your program for proper operation.

Program

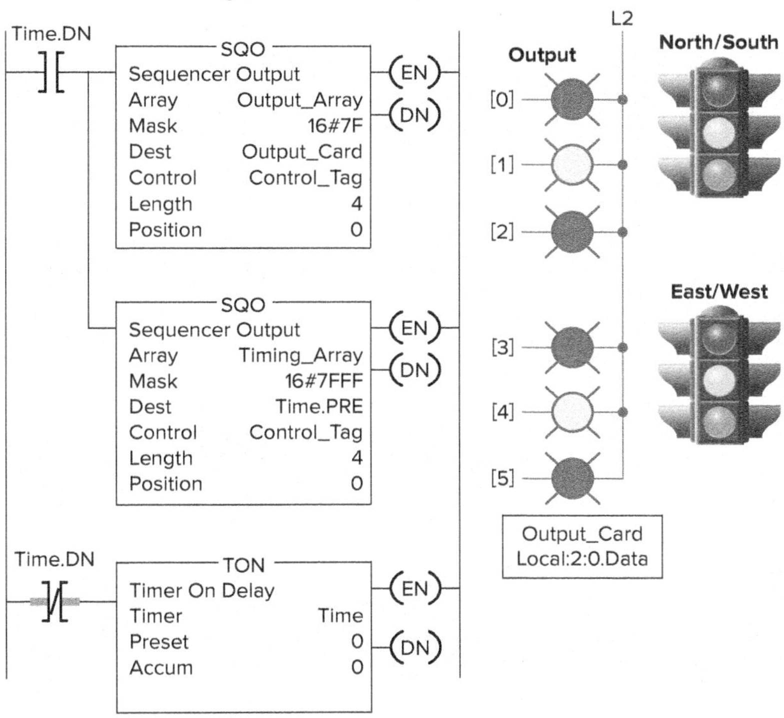

Timing chart

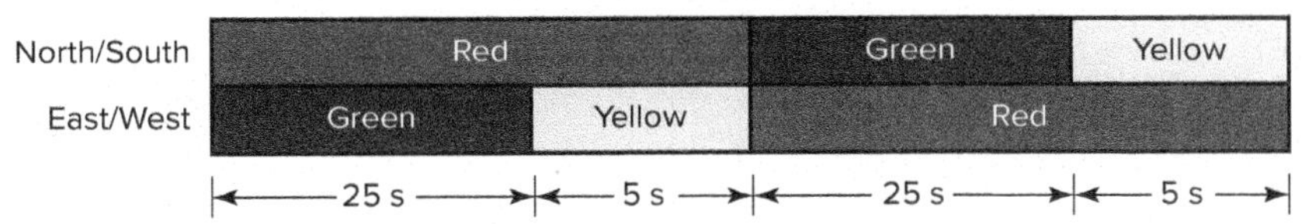

[-] Timing_Array	Tag Editor / Monitor Tags	Positions
[+] Timing_Array [0]	0	0
[+] Timing_Array [1]	25000	1
[+] Timing_Array [2]	5000	2
[+] Timing_Array [3]	25000	3
[+] Timing_Array [4]	5000	4

[-] Output_Array	Tag Editor / Monitor Tags	Positions
[+] Output_Array [0]	_0000_0000_0000	0
[+] Output_Array [1]	_0000_0010_0001	1
[+] Output_Array [2]	_0000_0001_0001	2
[+] Output_Array [3]	_0000_0000_1100	3
[+] Output_Array [4]	_0000_0000_1010	4

NAME ______________________________ DATE______________

12-6 Construct and download the time-driven constant interval sequencer program shown. Test your program for proper operation.

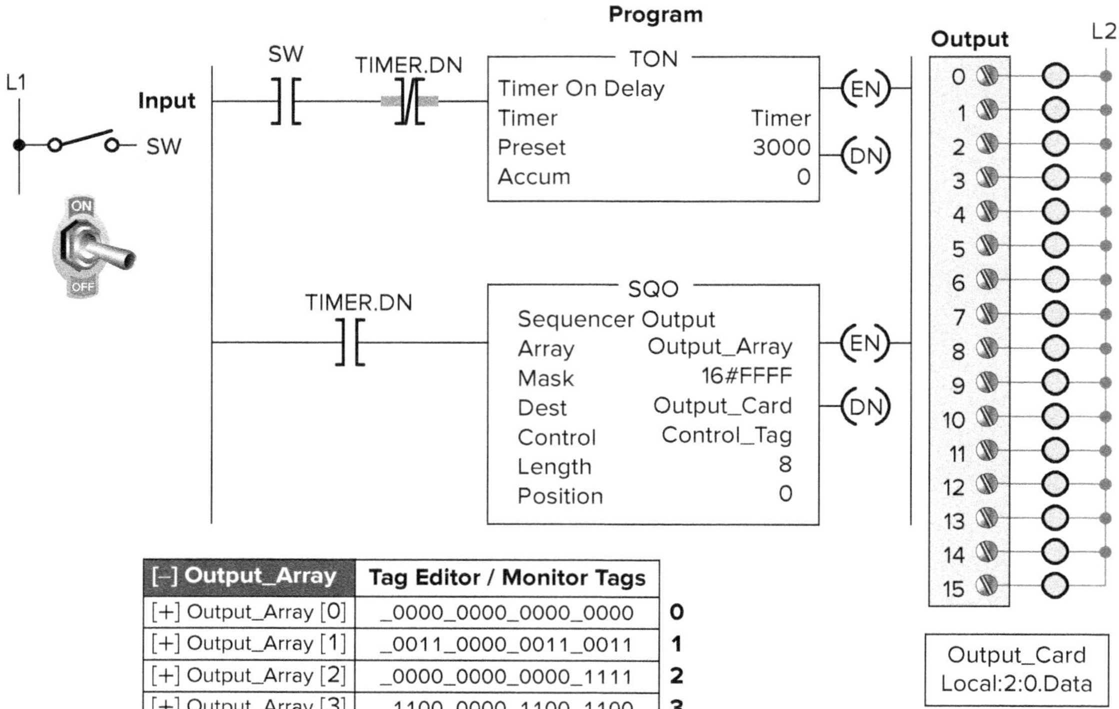

[–] Output_Array	Tag Editor / Monitor Tags	Positions
[+] Output_Array [0]	_0000_0000_0000_0000	0
[+] Output_Array [1]	_0011_0000_0011_0011	1
[+] Output_Array [2]	_0000_0000_0000_1111	2
[+] Output_Array [3]	_1100_0000_1100_1100	3
[+] Output_Array [4]	_0000_0000_1111_1111	4
[+] Output_Array [5]	_0000_0000_1100_0000	5
[+] Output_Array [6]	_0011_0000_1111_1111	6
[+] Output_Array [7]	_0001_0000_0001_1111	7
[+] Output_Array [8]	_0101_0000_0101_0101	8

NAME ____ DATE_______________

12-7 Construct and download the event-driven sequencer program shown. Test your program for proper operation.

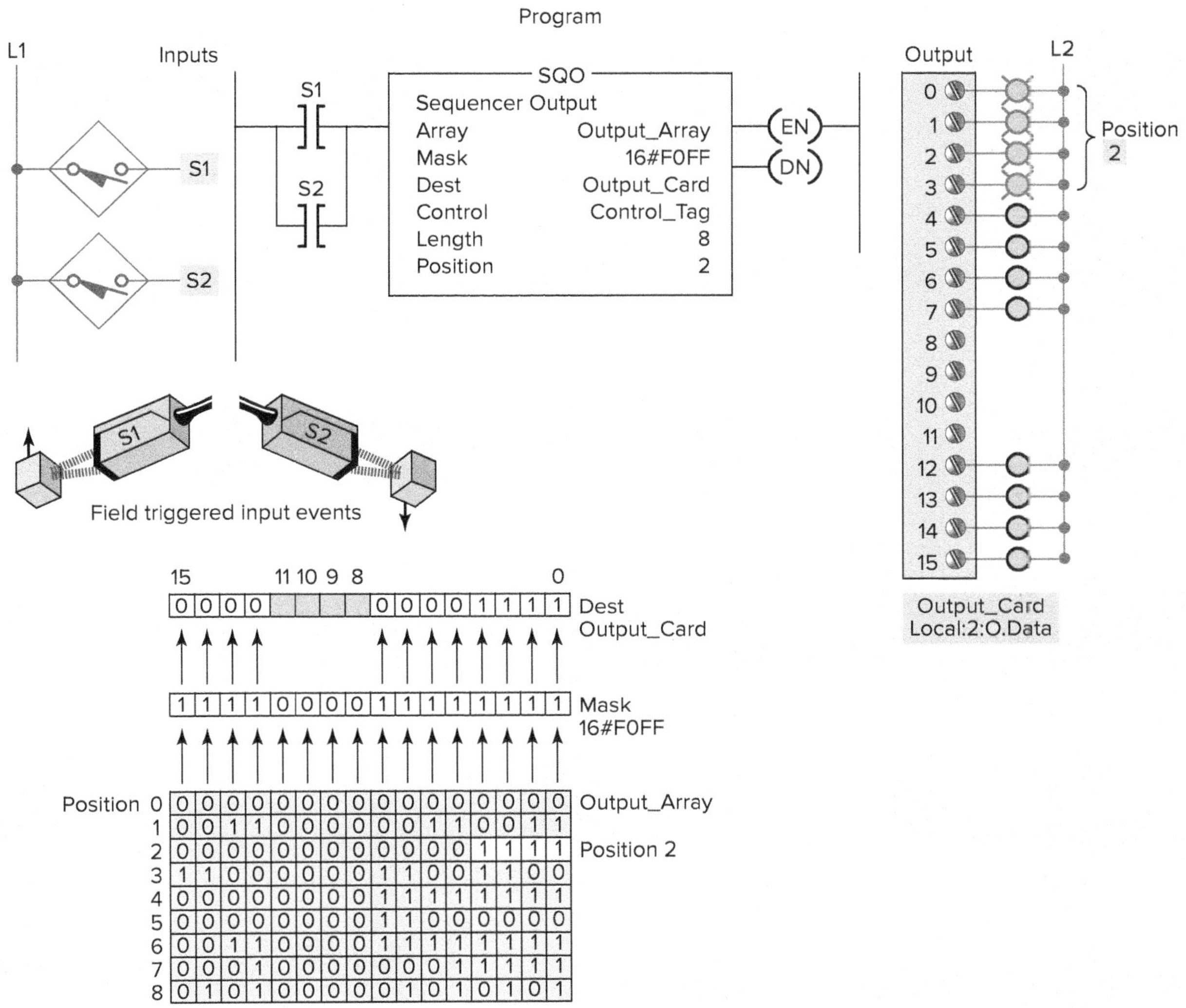

NAME ______________________________ DATE______________

12-8 Construct and download the sequencer load (SQL) instruction program shown. Test your program for proper operation.

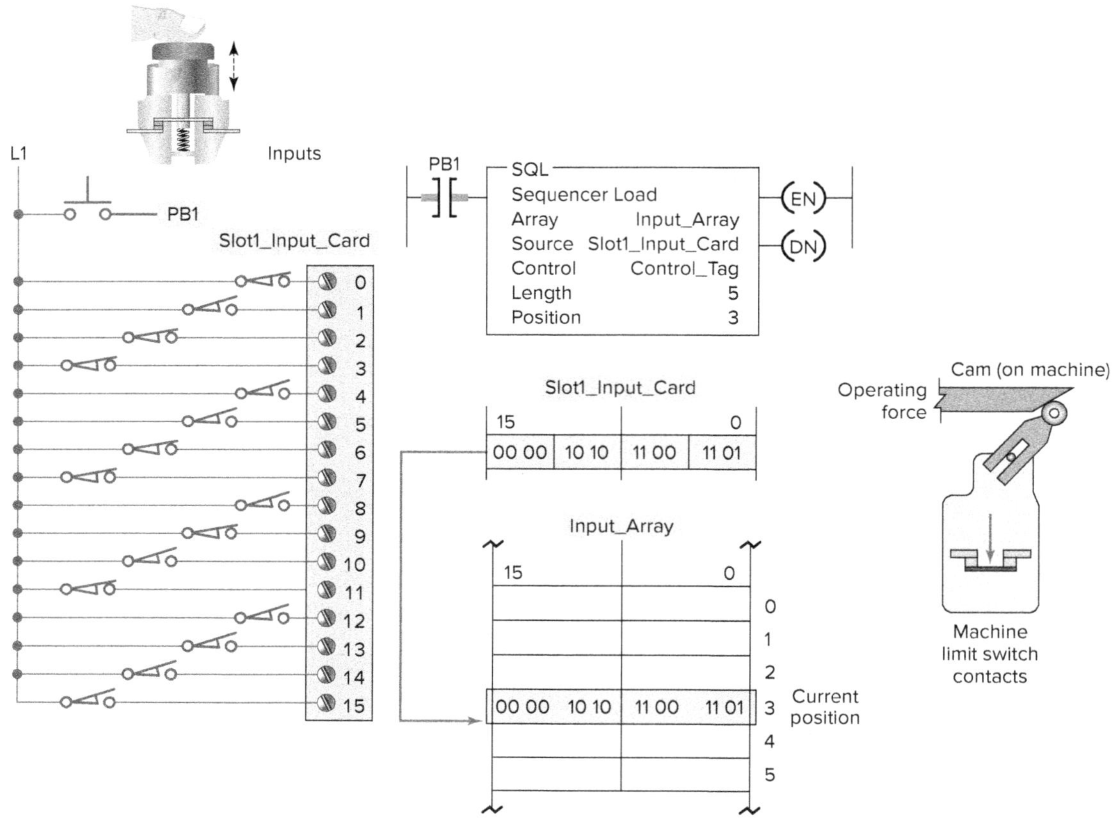

12-9 Construct and download the bit shift left (BSL) instruction program shown. Test your program for proper operation.

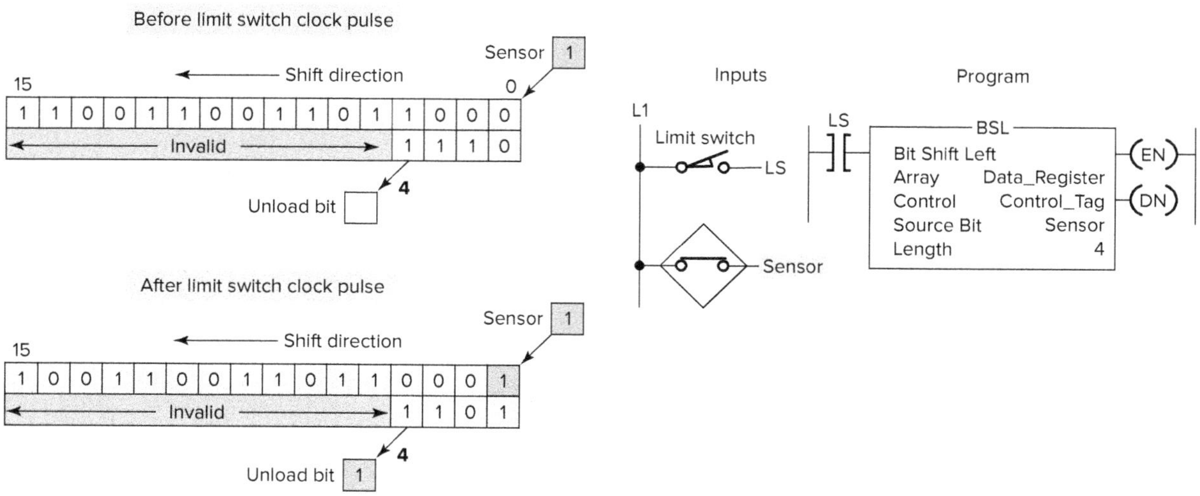

NAME ________________________________ DATE______________

12-10 Construct and download the bit shift right (BSR) instruction program shown. Test your program for proper operation.

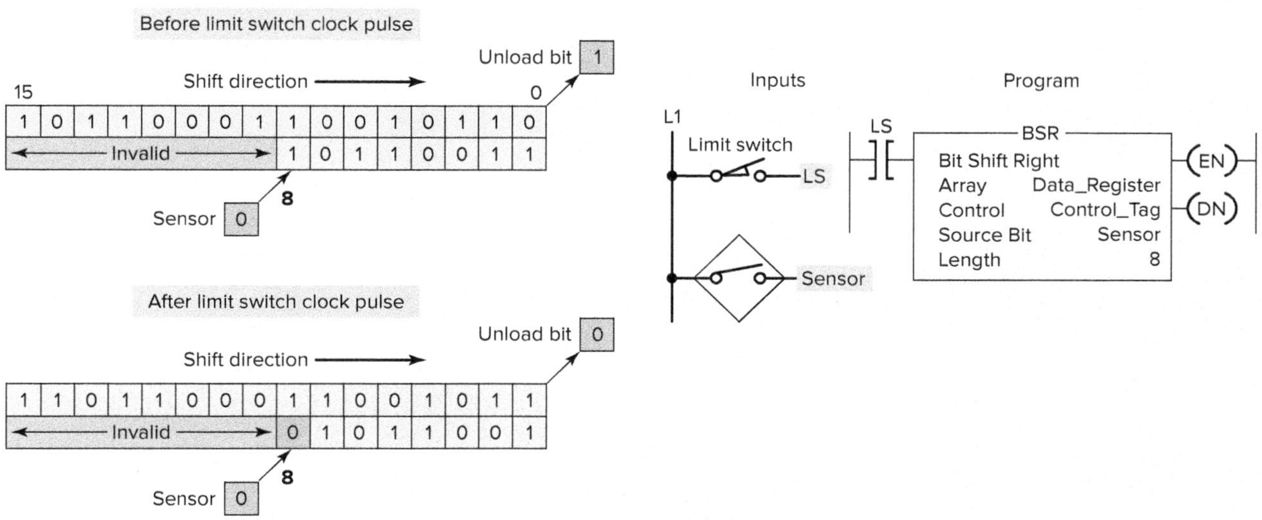

12-11 Construct and download the bit BSL instruction program with wraparound operation shown. Test your program for proper operation.

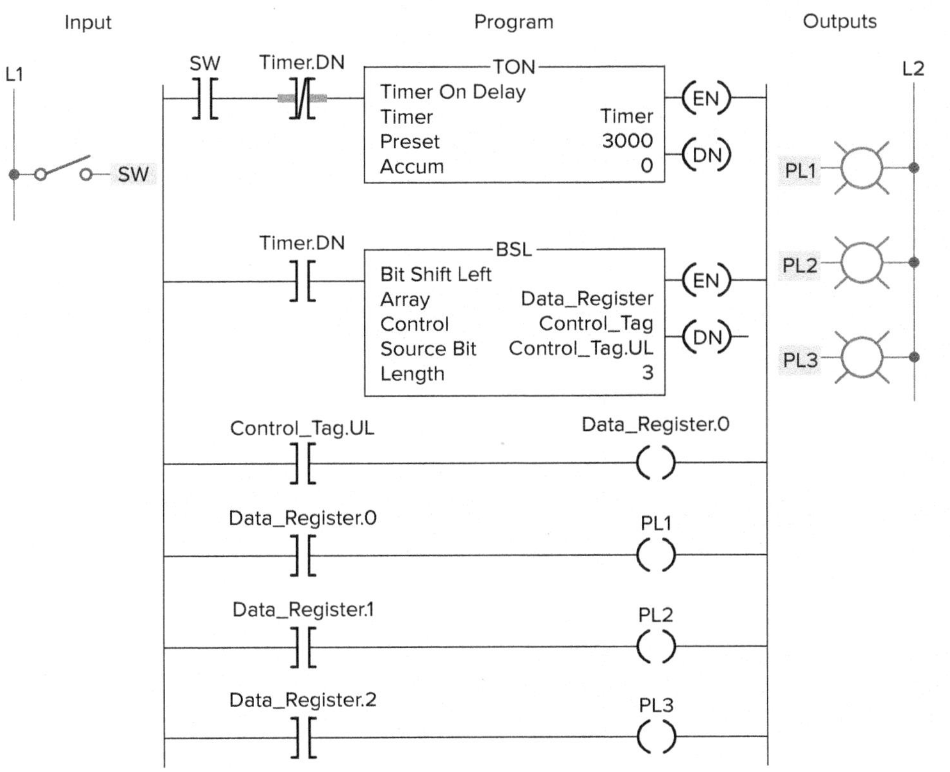

NAME ______________________________ DATE______________

12-12 Construct and download the spray-painting process program shown. Test your program for proper operation.

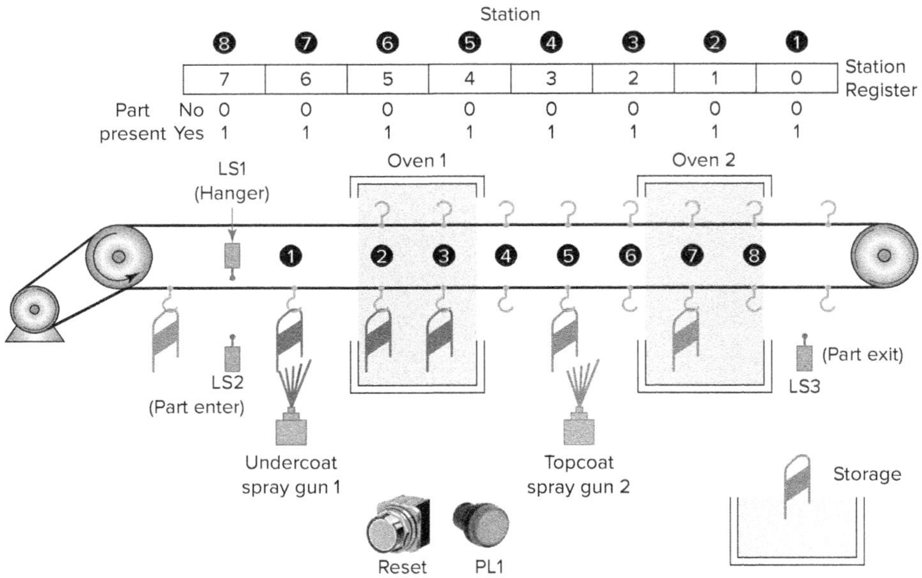

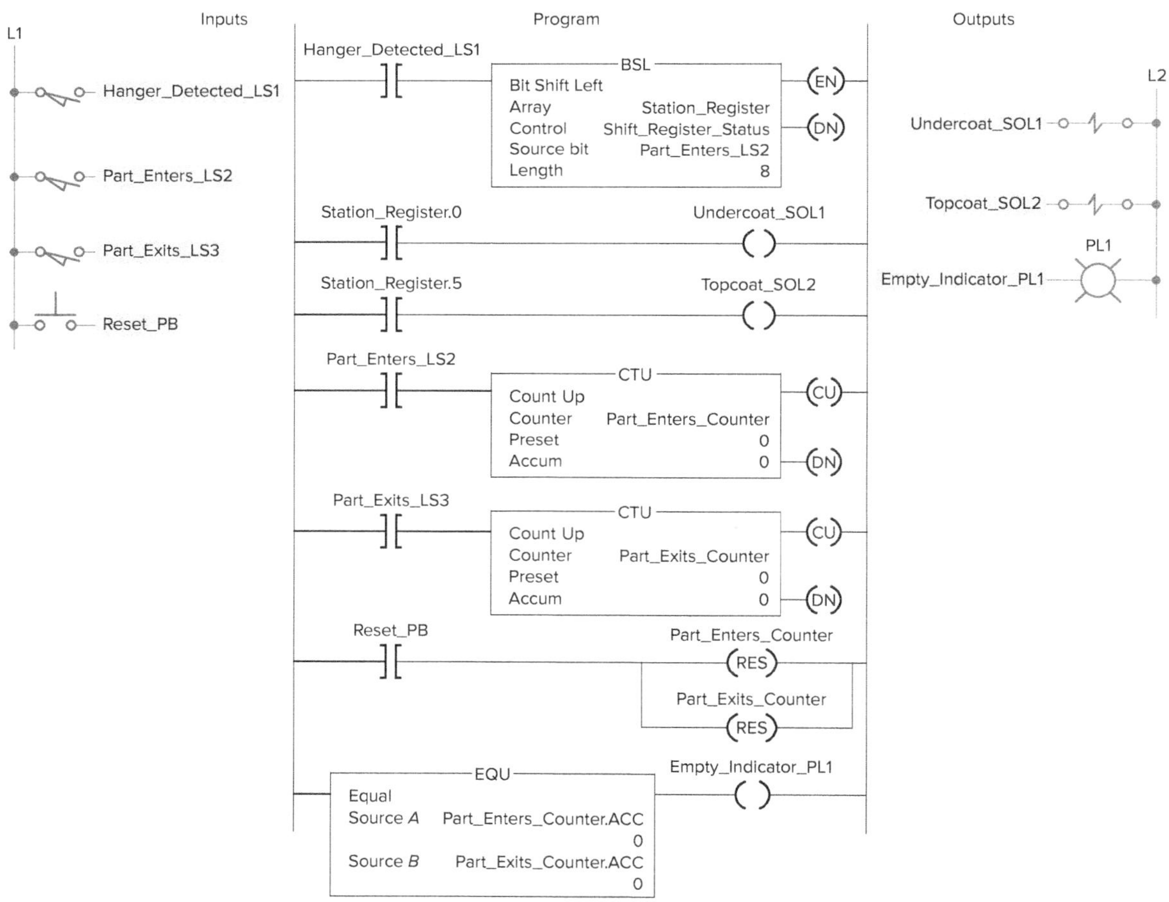

NAME ____ DATE________________

12-13 Construct and download the tracking of carriers flowing through a 16-station machine program shown. Test your program for proper operation.

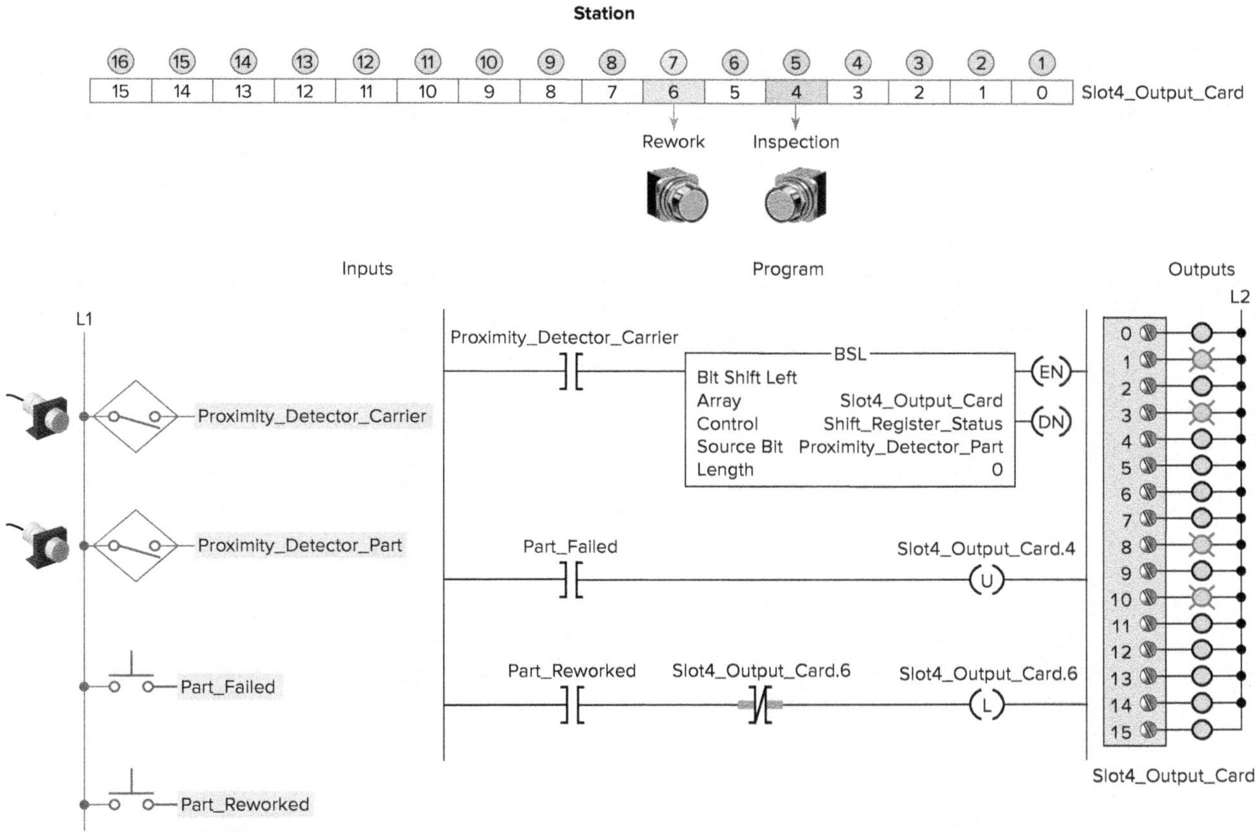

NAME ______________________________ DATE ______________

12-14 Construct and download the pass/fail inspection program shown. Test your program for proper operation.

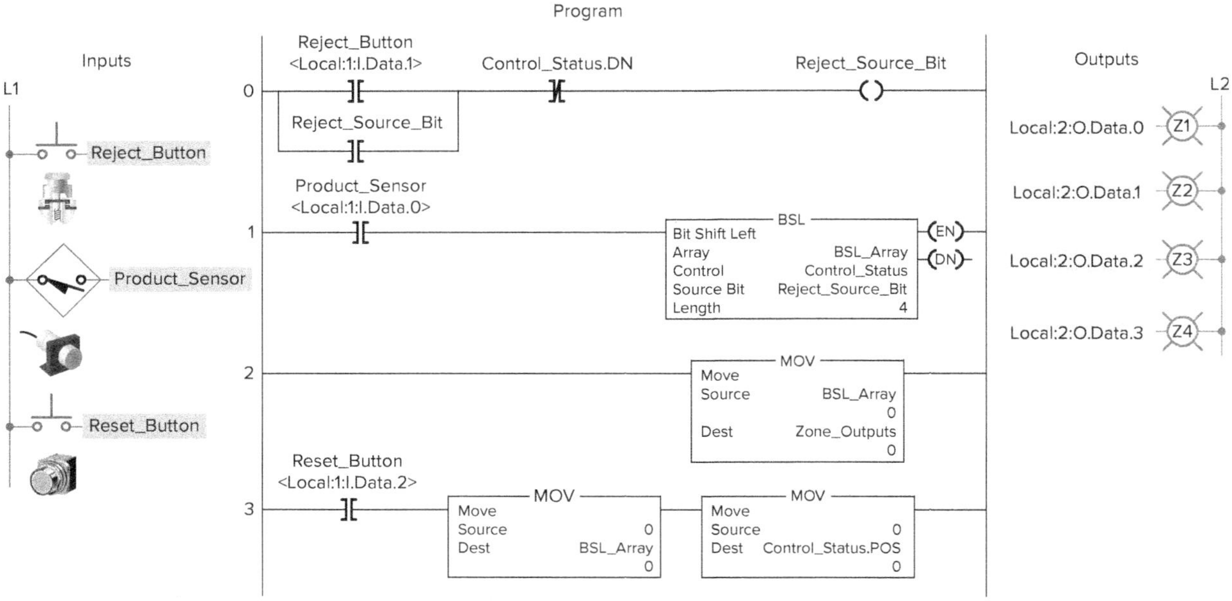

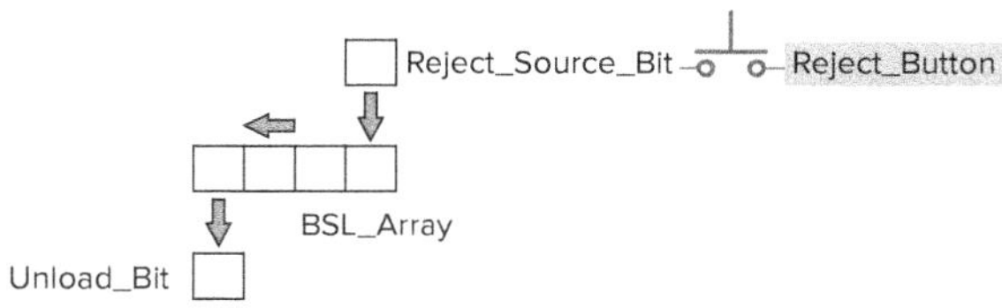

Tag Name	Value	Style	Data Type
⊞-BSL_Array	0	Decimal	DINT
⊞-Control_Status	{...}		CONTROL
⊟-Zone_Outputs	0	Decimal	DINT
Zone_Outputs.0	0	Decimal	BOOL
Zone_Outputs.1	0	Decimal	BOOL
Zone_Outputs.2	0	Decimal	BOOL
Zone_Outputs.3	0	Decimal	BOOL
Product_Sensor	0	Decimal	BOOL
Reject_Button	0	Decimal	BOOL
Reject_Source_Bit	0	Decimal	BOOL
Reset_Button	0	Decimal	BOOL

NAME ______________________ DATE ______________

12-15 Construct and download the data indexed in and out of a FIFO file program shown. Test your program for proper operation.

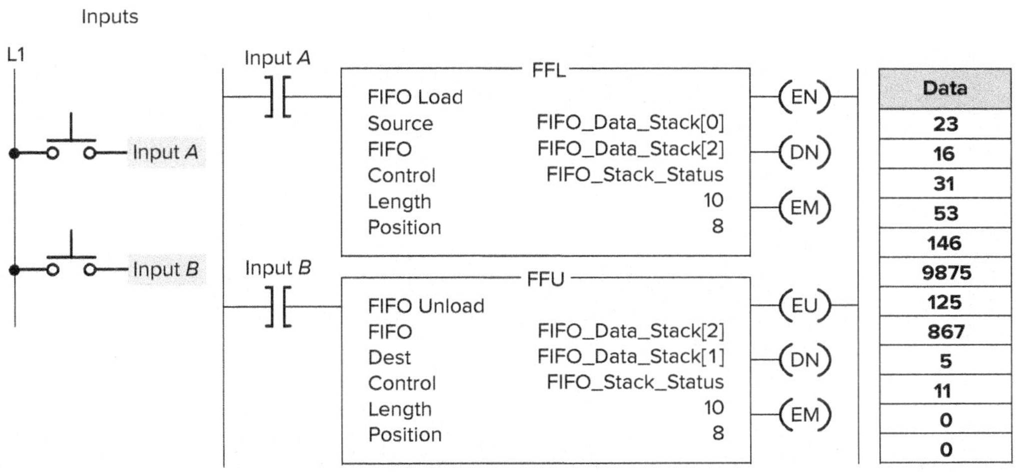

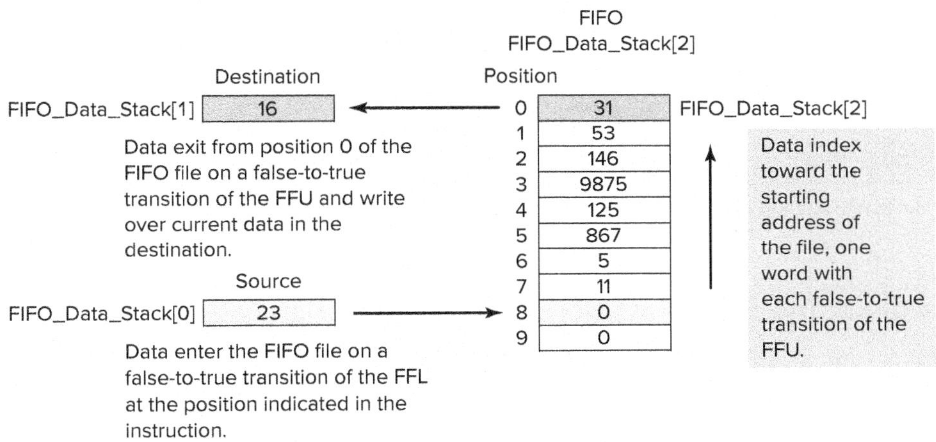

NAME ______________________________________ DATE________________

12-16 Construct and download the LIFO instruction pair program shown. Test your program for proper operation.

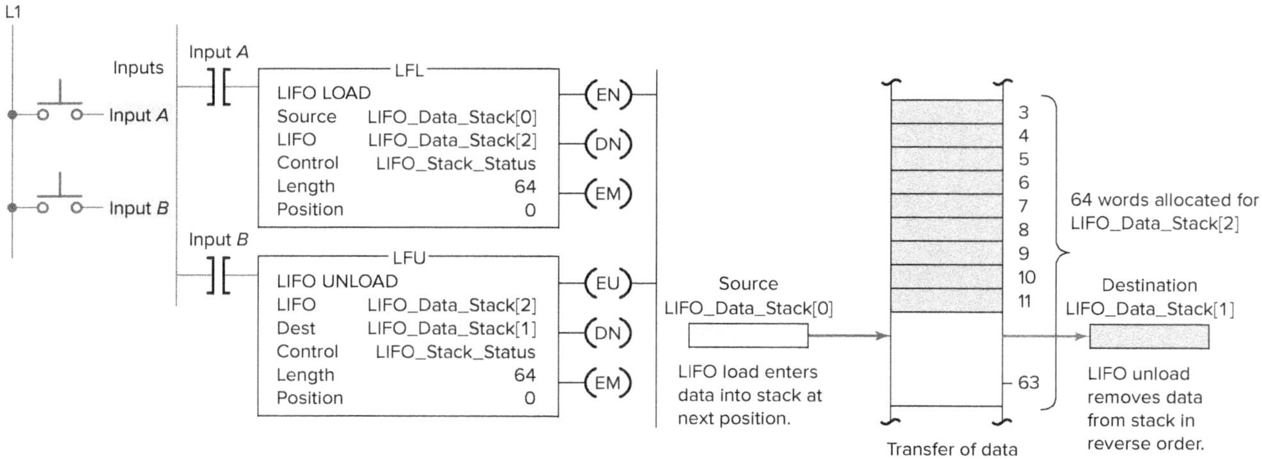

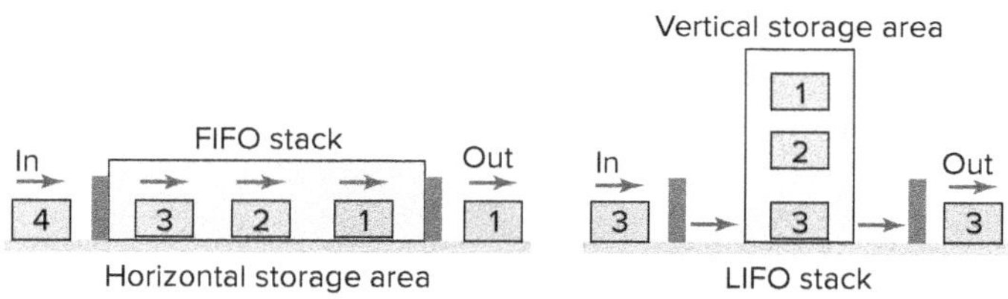

12-17 Construct and download the FIFO instruction as part of a data stack operation program shown. Test your program for proper operation.

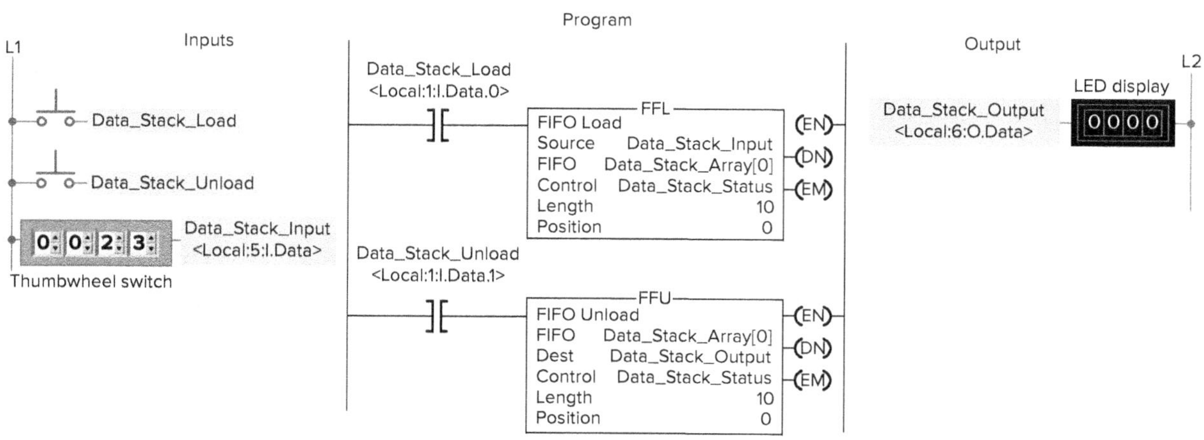

NAME ________________________________ DATE______________

12-18 Design and download a bit shift register program that will emulate the assembly line process described. Test your program for the required following operation:

- A product moves continuously down an assembly line that has four stations.
- The product enters the inspection zone, where its presence is sensed by a proximity switch.
- The inspector examines it and activates a reject button if the product fails inspection.
- If the product is defective, a reject status lights come on at stations 1, 2, and 3 to tell the assembler to ignore the part.
- When a defective part reaches station 4, a diverter gate is activated to direct that part to a reject bin.

12-19(a) Design, download, and test a PLC sequencer program which will operate as follows:

- The sequencer must step in 5-second intervals.
- Output No. 1 - on all the time the machine is cycling.
- Output No. 2 - on except for steps 3 and 5.
- Output No. 3 - on only in step 3
- Output No. 4 - on in steps 2 and 4.
- Output No. 5 - on in steps 2, 3, and 4.
- Output No. 6 - on in steps 1 and 5.

12-19(b) Modify the program to provide the following additional control features:

- All outputs must stay off and the sequencer must not operate until a start button is pressed.
- Once the start button is pressed, the sequencer completes one complete cycle and then stops automatically.
- Pushing a stop button resets and stops the sequencer.

NAME ______________________________ DATE______________

CHAPTER 13

PLC Installation Practices, Editing, and Troubleshooting

13-1 The program shown does not work as expected. Modify the program so that it operates correctly with the input and output devices shown. Download and test your program. Include a rung comment to describe what the original problem was and how you corrected it.

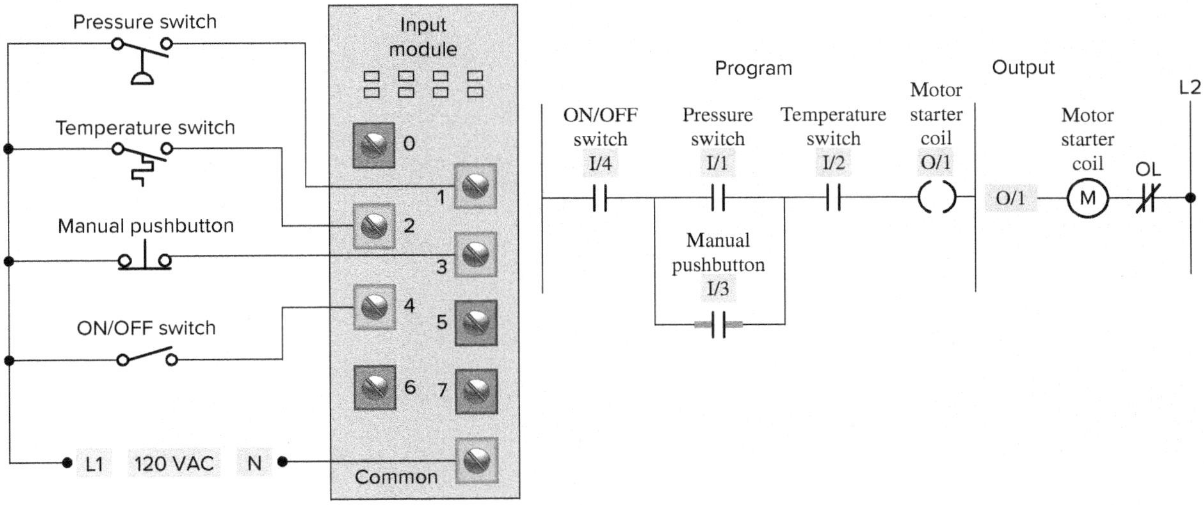

13-2 This three-wire control program is not working correctly. Modify the program so that it operates correctly with the input and output devices shown. Download and test your program. Include, as a rung comment, what the original problem was and how you corrected it.

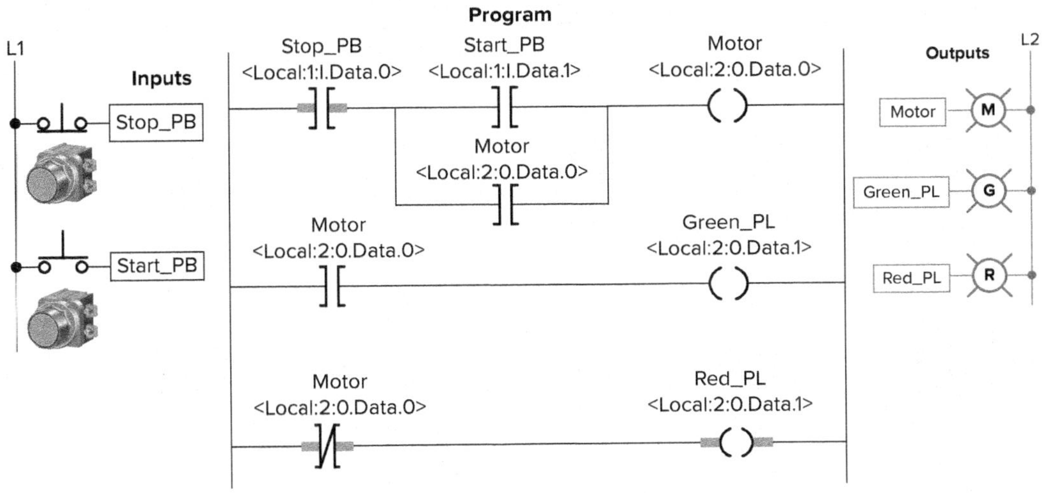

NAME ______________________________ DATE______________

13-3 Construct and test the program shown. Modify the program to compare a thumbwheel switch value of 6 with a constant value of 3. Download and test your program.

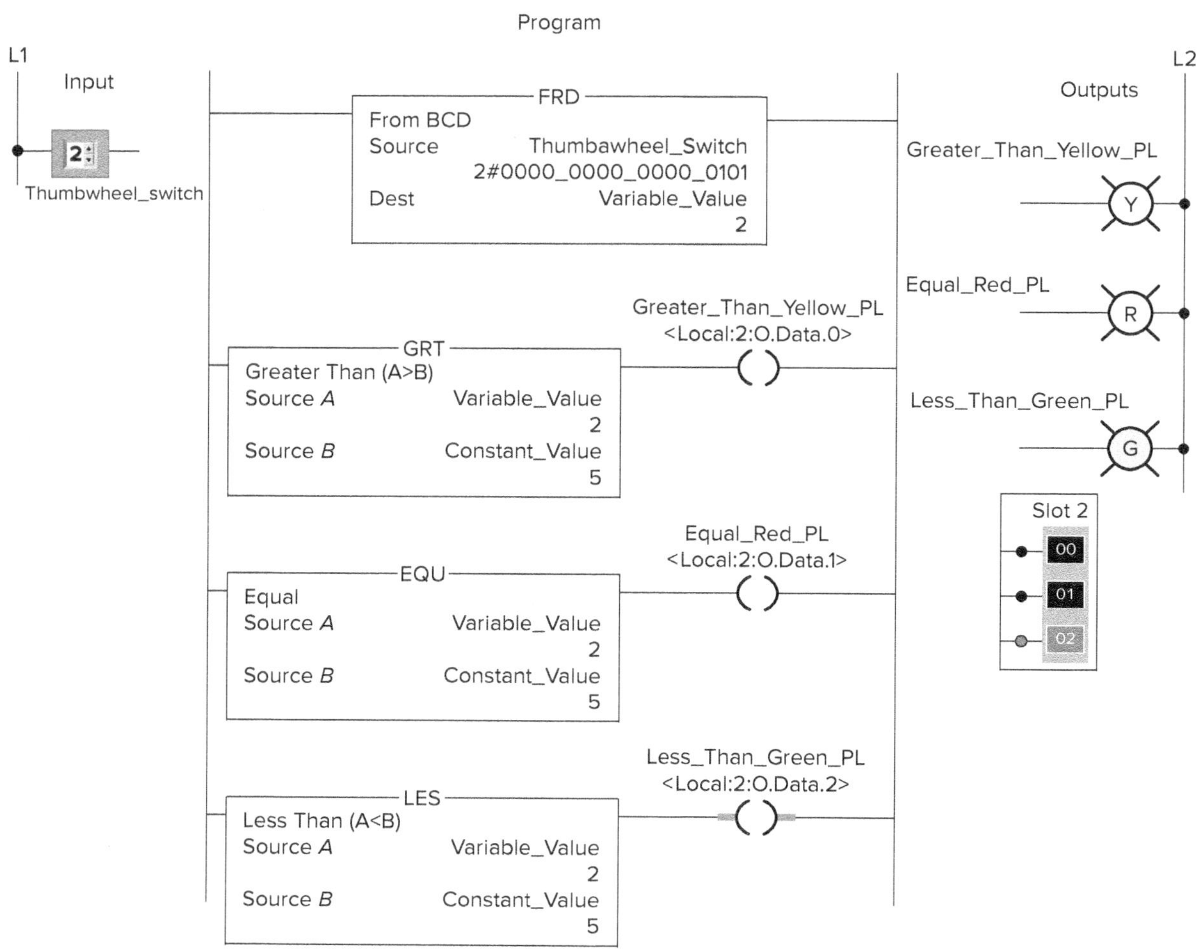

NAME ___________________________ DATE___________

13-4 The practical application of the program logic shown is to control a light from 3 positions using three single pole switches rather than two 3-way switches in combination with a single 4-way switch. Modify the program so that the light can be operated using four single pole switches.

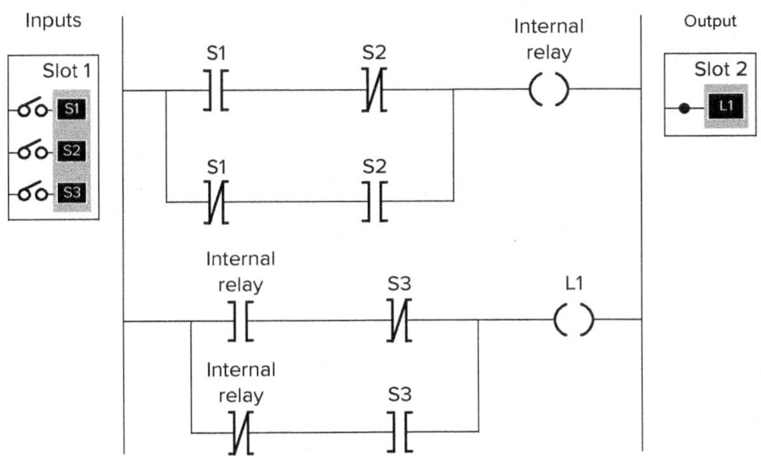

13-5 Reconstruct the PLC program shown using the following input pushbutton switches:

PB1 - Normally Closed Pushbutton
PB2 - Normally Open Pushbutton
PB3 - Normally Open Pushbutton
PB4 - Normally Closed Pushbutton
Download and test your program by verifying its operation with the use of a truth table.

NAME ______________________________ DATE______________

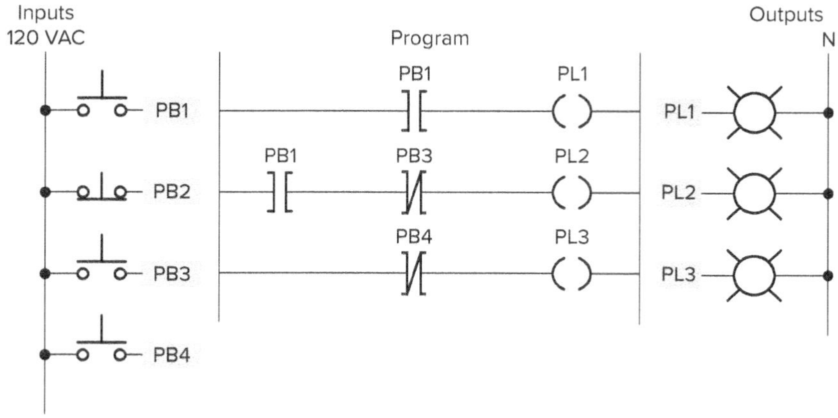

Truth Table

Input Device State	Input State				Output State		
	PB4	PB3	PB2	PB1	PL3	PL2	PL1
No PBs pressed	0	0	1	0	1	1	0
PB1 pressed	0	0	1	1	1	1	1
Pb3 pressed	0	1	1	0	1	0	0
PB2, 3, 4 pressed	1	1	0	0	0	0	0

13-6 The machine that uses the program shown has malfunctioned. The normally open Reverse button has been permanently damaged and needs to be replaced with a normally closed type. Reconstruct the PLC program to accommodate this change. Download and test your program for proper operation.

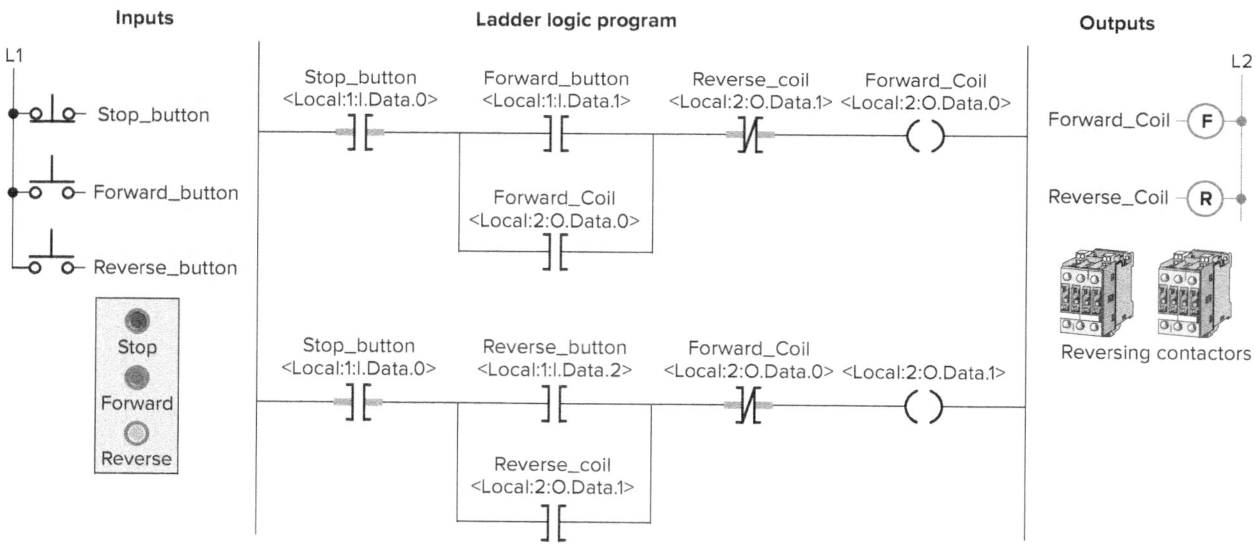

NAME ______________________________ DATE ____________

13-7 The program shown is not working properly.

The intended operation is:

- Motor M1 is started.
- Motor M2 is started automatically after the pressure switch closes.
- Motor M3 is started automatically 15 seconds after Motor M2 has started.

Examine the program and determine what error has occurred. Correct and test your program modification. Include, as a rung comment, what the original problem was and how you corrected it.

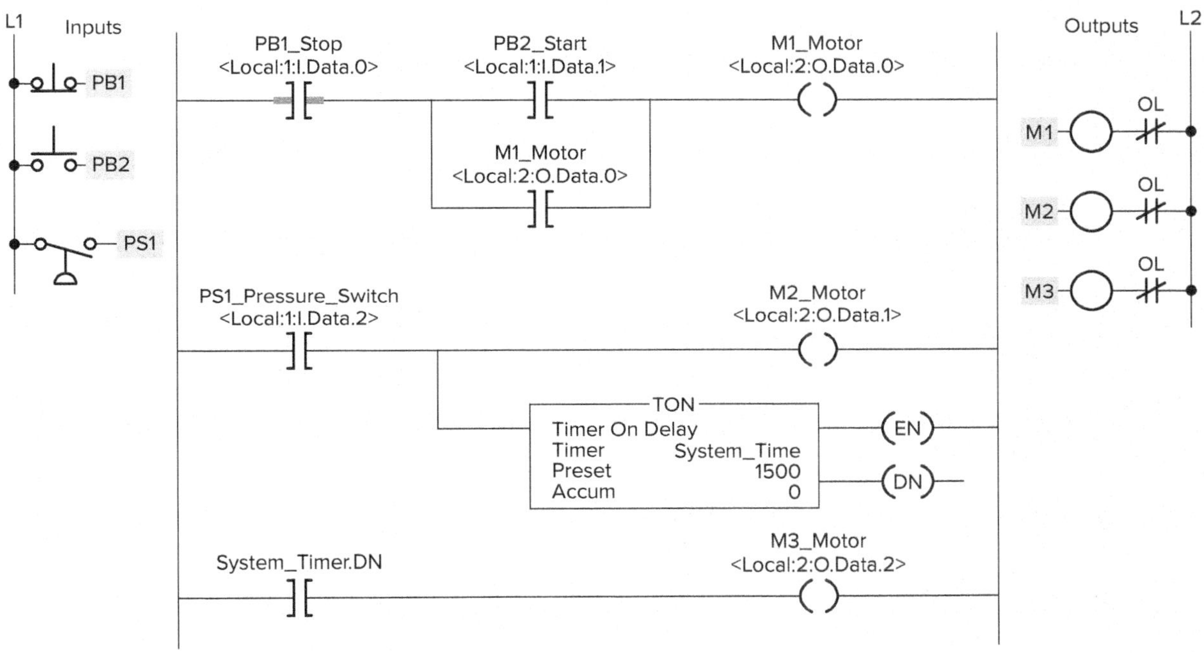

NAME ______________________________ DATE_______________

13-8 The program shown was downloaded and operated only to find that the motor failed to start when the Start_PB button was pressed. Examine the program and determine the cause. Correct and test your program modification. Include, as a rung comment, what the original problem was and how you corrected it.

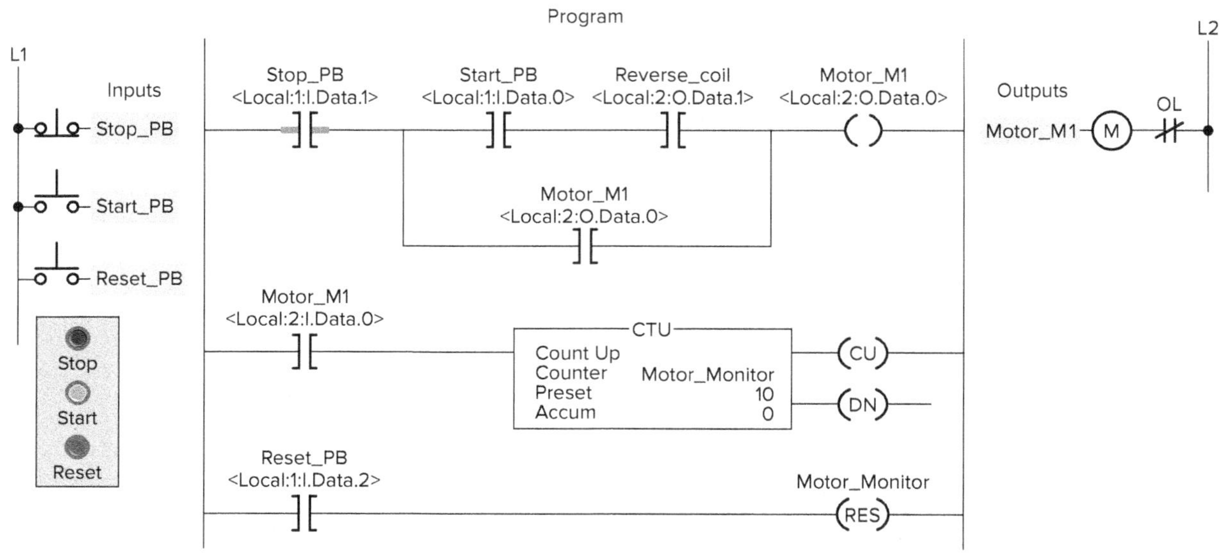

13-9 The program shown was downloaded to the PLC but did not execute properly. Examine the program and determine the cause. Correct and test your program modification. Include, as a rung comment, what the original problem was and how you corrected it.

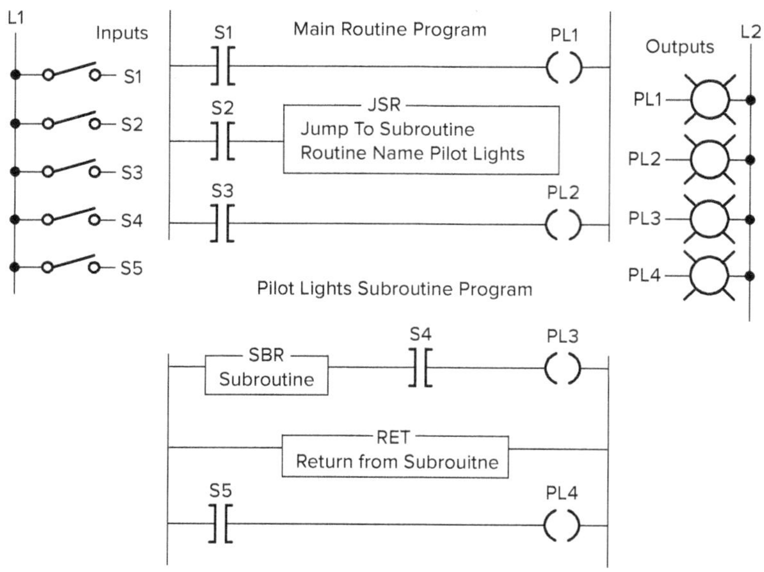

NAME ______________________________________ DATE______________

13-10 The ladder logic shown below is for a Limit Test instruction. Modify the program shown so that PL1 is on any time the accumulated value of the counter ranges between 18 and 36. Download and test your program.

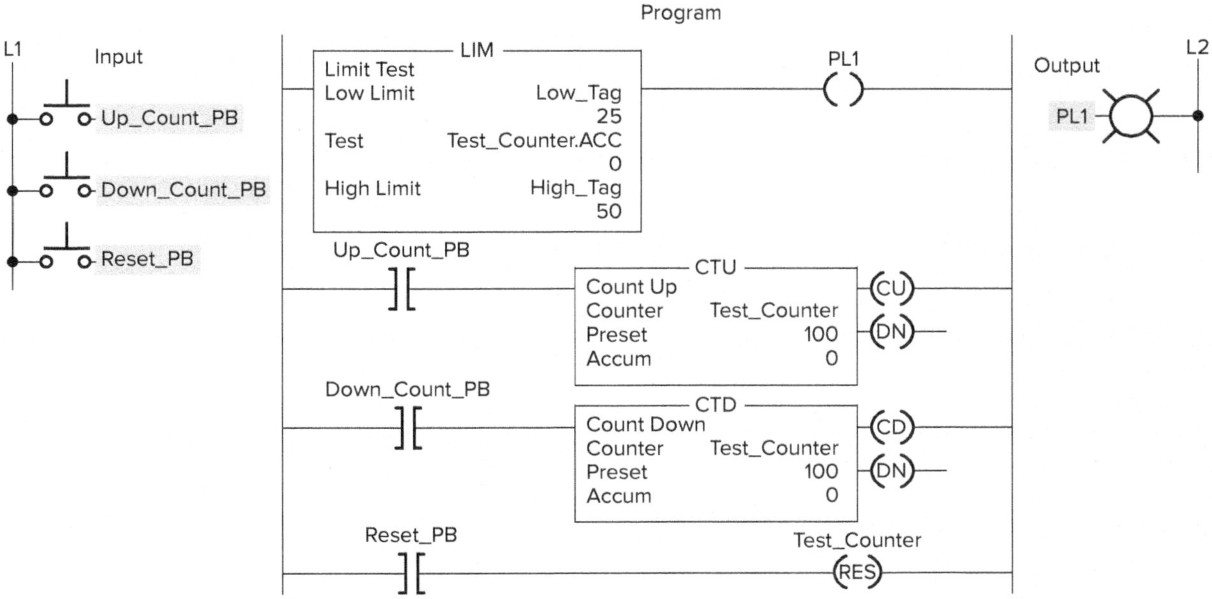

NAME ______________________________ DATE ______________

13-11 This assignment is in reference to a time-driven sequencer. The program shown was downloaded to the PLC but did not execute properly. Examine the program and determine the cause. Correct and test your program modification. Include, as a rung comment, what the original problem was and how you corrected it.

Program

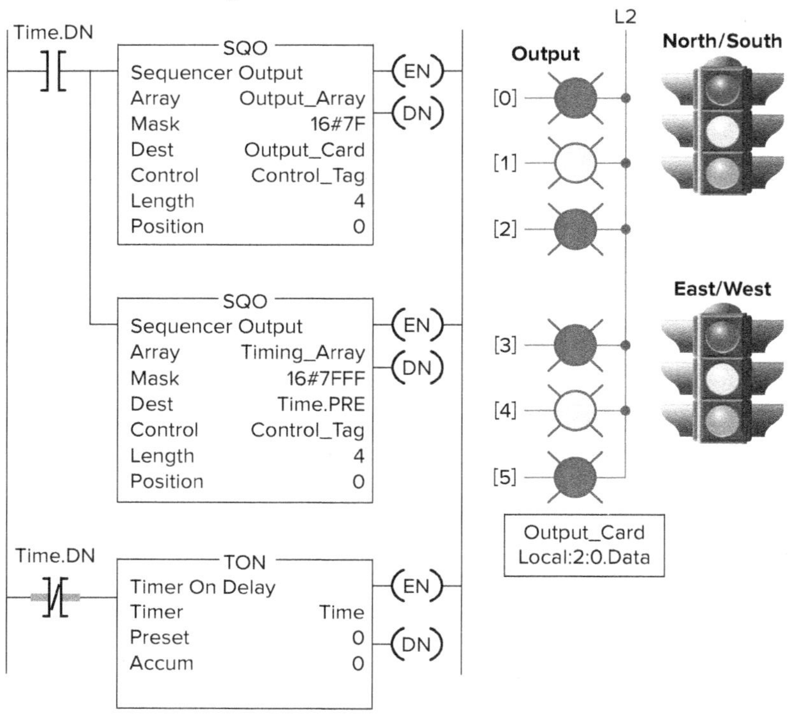

Timing chart

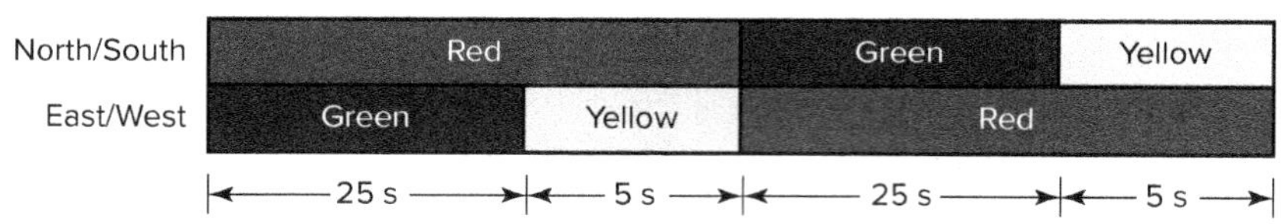

[–] Timing_Array	Tag Editor / Monitor Tags	Positions
[+] Timing_Array [0]	0	0
[+] Timing_Array [1]	5000	1
[+] Timing_Array [2]	25000	2
[+] Timing_Array [3]	5000	3
[+] Timing_Array [4]	25000	4

[–] Output_Array	Tag Editor / Monitor Tags	Positions
[+] Output_Array [0]	_0000_0000_0000	0
[+] Output_Array [1]	_0000_0010_0001	1
[+] Output_Array [2]	_0000_0001_0001	2
[+] Output_Array [3]	_0000_0000_1100	3
[+] Output_Array [4]	_0000_0000_1010	4

